Toyotas

Motogard
CLASS
4
TOYOTA TEAM EUROPE
PIRELLI
BILSTEIN
DACHSER
TOYOTA

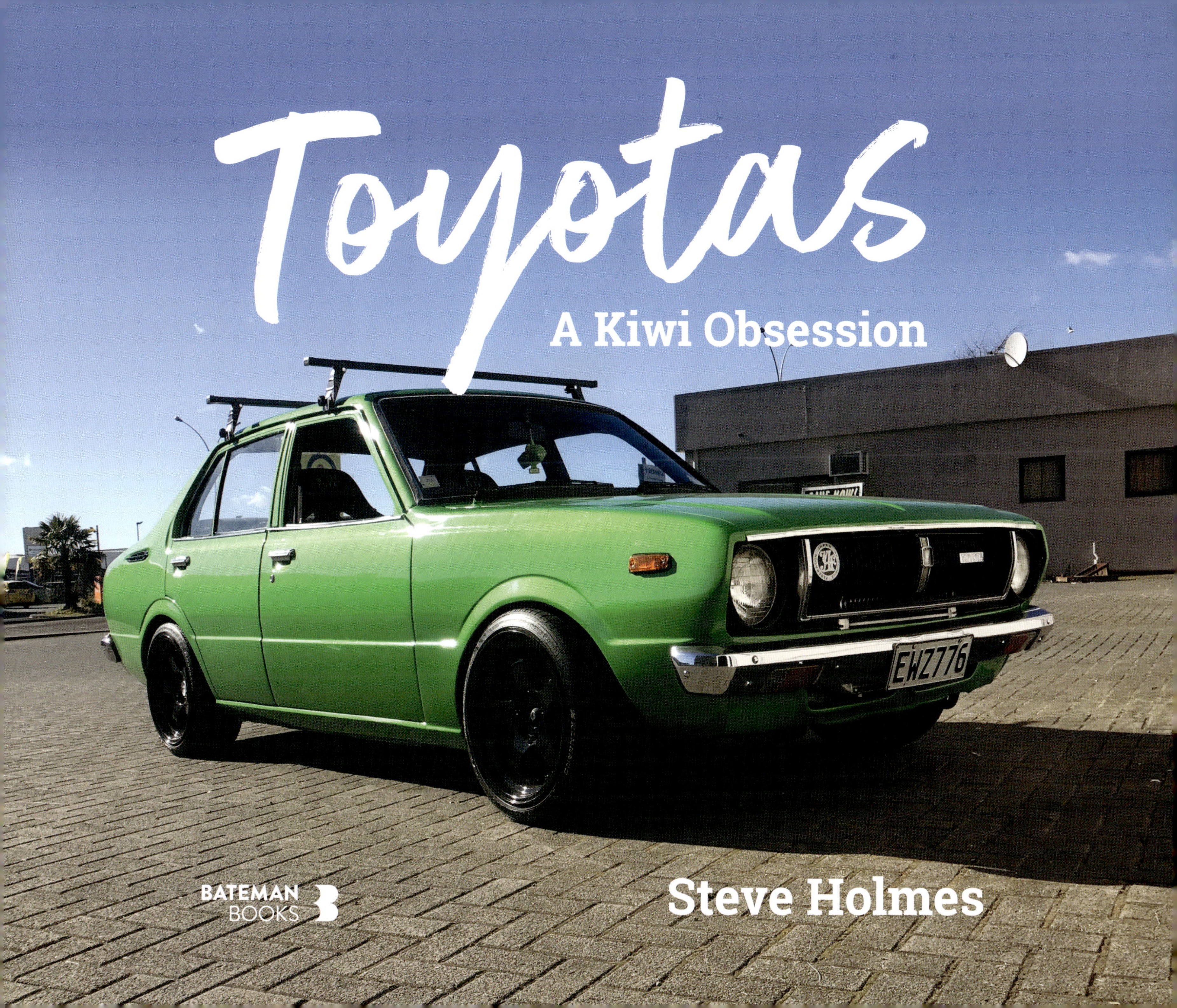
Toyotas
A Kiwi Obsession
EWZ776
BATEMAN BOOKS
Steve Holmes

Published in 2022 by David Bateman Ltd
Unit 2/5 Workspace Drive, Hobsonville, Auckland 0618, New Zealand
www.batemanbooks.co.nz

ISBN 978-1-98-853815-0

Imagery: As listed on page 230
Book design: Cheryl Smith, Macarn Design
Printed in China through Asia Pacific Offset Ltd

Contents

Introduction

Like many long-standing motor vehicle manufacturers, Toyota got its start producing products outside the automotive industry. In 1924, Sakichi Toyoda invented the Toyoda Model G Loom, quite possibly the most advanced loom in the world at the time. The Toyoda Automatic Loom Works machine was designed in such a way that a single unskilled worker could supervise 30 to 50 looms at a time, thus enabling mass-production.

In 1929, licensing for the automatic loom was sold to the Platt Brothers company in Britain, generating significant cash flow for Toyoda, that at least in part, was used to establish Toyoda Motor Corporation in 1933. Under the direction of Kiichiro Toyoda, the company spent the next two years producing various prototypes, including an engine, a truck, and a car. The first Toyoda Motor Corporation vehicle to go into production was the G1 truck, modelled on similar products produced by Ford and Chevrolet, with a sales price slightly cheaper than the American alternatives.

In 1936, the first passenger car was launched; the Toyoda Model AA. Much like the G1 truck, the AA sedan borrowed heavily from vehicles being produced by Detroit, with styling similar to the Chrysler Airflow. Construction of the bodies was outsourced by Toyoda, who then placed them on their own chassis. Propulsion was from Toyoda's own 3389cc inline-6 motor, itself a copy of the engines found in Chevrolets. Again, the AA was specifically aimed to be cheaper than its American rivals on the Japanese market.

While the automotive company initially carried the family name Toyoda (トヨダ), it was Rizaburo Toyoda, son-in-law of Sakichi Toyoda, who suggested a subtle name change, to Toyota (トヨタ), because it required eight brush strokes to write in Japanese, and in Japanese culture, eight is a lucky number. Furthermore, Toyota has a voiceless consonant, which makes for a clearer sound, whereas Toyoda has a voiced consonant, resulting in it being slightly more 'muddy'. In addition, being a technically advanced company, it wanted to distance itself from any association with farming; an unfashionable industry in Japan. The name Toyoda literally means 'fertile rice paddies'.

Throughout the late 1930s, Toyota Motor Corporation grew quickly, thanks in part to the Japanese government blocking the importation of American vehicles, thus allowing it a less competitive market in which to operate.

With the onset of World War II, Toyota was instructed to aid Japan's war efforts. It was while occupying the Philippines that the Japanese Imperial Army got hold of an American Willys MB Jeep, and sent it back to Japan. Toyota was then ordered to produce a similar vehicle, but with altered appearance, resulting in the Model AK. There followed the AK10, a light truck requested by the Japanese government for the military, designed by reverse-engineering a Bantam GP, and powered by a Toyota Type C 4-cylinder engine.

Like many Japanese manufacturers, Toyota Motor Corporation was greatly impacted by World War II, and its Koromo Plant was bombed just one day before Japan surrendered.

Following the war, and under the US-led occupation, Japanese manufacturers were banned from producing passenger vehicles. Toyota was, however, given approval to build trucks for passenger use.

In 1949, production of passenger vehicles was again granted, but the Japanese economy had tanked. Toyota was forced to dramatically downscale, closing plants and laying off workers. It was later aided in its recovery thanks to an order for 1000 trucks from the

US government as part of its involvement in the Korean War.

The Toyota manufacturing process underwent wholesale changes following a trip by company executives to Ford Motor Company in 1950. There they were trained in the 'just in time system' that Ford had been perfecting for decades to improve and speed up production, and to reduce cost and wastage. The just in time system, or what would later become known as lean manufacturing, involves having just the right number of parts supplied at just the right time to be implemented into the assembly process, thus improving productivity and costs. This was incorporated by Toyota and further developed into its famed Toyota Production System.

In 1952, Toyota began design of its very first fully-fledged passenger cars; the Toyopet Crown and Master. The Crown was intended as personal transport, while the Master was designed as a commercial vehicle, largely for use as a taxi. For the first time, everything would be made in-house by Toyota Motor Corporation. After three years of development, during which time the company was struck by the sudden death of Kiichiro Toyoda, the Toyopet Crown and Master went on sale in 1955.

Meanwhile, having produced a new AK10 all-terrain vehicle (the 'Jeep' BJ), Toyota came under pressure from Willys Company for claimed trademark violation, and thus rebranded it the Land Cruiser. In 1951, Toyota's test driver Ichiro Taira had driven a Jeep BJ up to the Stage 6 (of 10) of the 3776-metre-high Mt Fuji, becoming the first vehicle to accomplish the feat. The National Police Agency were on-hand to observe the achievement, and promptly ordered nearly 300 examples, which became its official patrol vehicle.

In 1955, Toyota began export of the Land Cruiser in CKD (completely knocked down) kit form to Latin America. There followed further exports to Burma, the Philippines, Thailand, and Australia. In 1958, a new manufacturing plant was constructed in Brazil. This was the first Toyota manufacturing plant outside Japan.

Toyota introduced its latest passenger vehicle, the Corona, in 1957. It was positioned beneath the Crown in Toyota's hierarchy system, and was slightly smaller and less powerful, with more conservative styling.

That same year, the Toyopet Crown was exported to the United States, but was poorly received, with only a handful of cars being sold. The robust little brute was designed for rough Japanese domestic roads, but in the US, it demonstrated a lack of power, with styling that polarized opinion, and was deemed heavily over-priced for what it was. Two years later, it was quietly withdrawn.

The early 1960s saw the Japanese economy go from strength to strength, and with it came a notable increase in public spending money. The Japanese government invested heavily in massively upgrading the nation's roading system, and customers flocked to car dealership showrooms to begin experiencing the freedoms that modern motoring offered.

Toyota re-entered the United States market with its new third-generation Corona in 1965. Unlike its failed attempt with the Crown, the new Corona was designed from the ground-up to appeal to American buyers, and was far better received. By 1967, Toyota was the third best selling import brand in the US, with more than 20,000 vehicles sold.

Meanwhile, back home, the strong local economy spurred an enthusiasm for the motor vehicle not previously seen in Japan. Motor racing became a popular form of entertainment, and manufacturers were soon involving themselves in the rigours of competition to promote the sporting attributes of their products. It also prompted manufacturers to start creating 'halo' cars; personal sports and luxury cars created not specifically as a mode of transportation, but instead, purely for enjoyment. In 1965, Toyota unveiled its exquisite 2000GT sports car, which went on sale two years later. Only about 350 of them were built. In today's international classic car market, Toyota 2000GTs can fetch over US$1 million, on the rare occasion they come up for sale.

In 1966, Toyota launched the Corolla as its entry into the new compact market, to compete against the Datsun 1000 and Subaru 1000.

Where Toyota really got a foothold in the US market was with the advent of the global energy crisis in the early 1970s. Suddenly, colossal gas-guzzling American V8s became highly undesirable, as buyers beat a hasty path to the various Japanese showrooms, including those of Toyota and Datsun, to acquire an economical compact. Having then enjoyed the value for money, build quality, and superior customer service these brands offered, many customers simply kept buying Japanese.

Throughout the 1990s and 2000s, while domestic American brands shut down many of their plants and moved to South America to take advantage of cheaper manufacturing costs, Toyota actually expanded its manufacturing in the US, building new plants, and extensively growing its work force. In 2021, the Toyota RAV4 was the best-selling vehicle in the US car and SUV market (selling 407,739 units), with the Camry third, the Highlander fifth, and the Corolla eighth. In the all-important light truck market, which is still larger than the car market (top seller Ford sold 726,000 F-Series pick-ups in 2021), Toyota sits fourth, and first non-domestic brand.

One of Toyota's strong suits has always been its engines, which remain popular among tuners, and are both strong, and have a willingness to produce power. Toyota partnered with Yamaha to create some of the greatest engines in automotive history, including the twin-cam inline-6 that propelled the 2000GT, from which various incarnations were developed.

Toyota, like all the Japanese manufacturers looking to establish themselves in the US market, used motor racing as a way of achieving this. In 1968, the company contracted Shelby American, headed by racing legend Carroll Shelby, to oversee a competition programme with its 2000GT in SCCA (Sports Car Club of America) C/Production sports car racing. Throughout the 1970s and '80s Toyota continued to expand on its racing involvement in the United States, competing in a variety of disciplines, from sedan competition right through to prototype sports car racing.

The 1990s saw Toyota partner with another American racing legend in Dan Gurney, to compete in the Indycar open wheeler racing series. In 2000, Toyota entered NASCAR (National Association for Stock Car Auto Racing), with an initial campaign in the Goody's Dash, before making the step up to the second-tier Xfinity Series where it spent a few years working with top teams and establishing itself until finally making the move to the top-tier NASCAR Cup Series. In this, the most popular and richest form of racing in the United States, Toyota has enjoyed countless championship and race wins, including at the highly celebrated Daytona 500.

Toyota is now a global brand, selling in most markets throughout the world. While its products are renowned for their reliability and ease of use, at its core are genuine enthusiasts who love cars, and love to create excitement. Toyota has amassed an armoury of enthusiast driving machines throughout the decades, and has retained a constant motorsport involvement, from domestic touring car and sports car racing, to the World Rally Championship, Formula 1, and the Le Mans 24 Hours, of which it can now rightly claim to be a multiple outright winner.

The Toyota Crown is the longest continuously running vehicle name plate in the automotive world, and the Corolla is the best-selling, having long-since surpassed the Volkswagen Beetle.

Toyota In New Zealand

Toyota entered the New Zealand market in the mid-1960s, through Consolidated Motor Distributors, who imported fully assembled cars from Japan.

Consolidated Motor Distributors later arranged to have Completely Knocked Down (CKD) cars assembled by Steel Motor Assemblies in Christchurch, and Campbell Motor Industries in Thames. The first car assembled in Christchurch was a Corona, in

1967. The first car assembled in Thames was a Corolla, in 1968.

Toyota, much like its Japanese counterparts who also made their way into the New Zealand market during the 1960s, produced modest, compact economy cars, which were well built and competitively priced. But Japanese cars were still a relatively new concept in New Zealand, whose history was deeply rooted in American, British and European products. While Australian cars from Ford and Holden were also relatively new to the local market, they were quickly accepted, being well-suited to NZ road conditions.

The Japanese brands had a harder time of it, but through building a reputation for quality, reliability, and competitive pricing, they forged their way into the New Zealand landscape, and are now the most popular in the domestic market. In 2021, Toyota vehicles finished second and fourth (RAV4 and Corolla respectively) in the new car/SUV market, and second and sixth (Hilux and HiAce) in the commercial vehicles market.

During the 1980s, Toyota developed a partnership with New Zealand racing driver Chris Amon who, during the 1960s and '70s, competed at the very highest level of the sport, driving for Ferrari in Formula 1 and sports car racing, while sharing the winning Ford GT40 with fellow countryman Bruce McLaren to give Ford its first Le Mans 24 Hours victory in 1966. It was through Amon that various Toyota products were developed specifically for New Zealand's unique roading demands, and the Amon Corona was perhaps the very first signature car seen in this country.

For Toyota in particular, one avenue for proving the value of its products was achieved through motorsport competition. In the early days, it ventured into rallying. Back in the late 1960s and throughout

the 1970s, rally events were brutal endurance competitions, spanning several days, and featuring both day and night stages, and invariably staged throughout the harsh winter months.

But Toyota has remained perhaps the most constant brand of any manufacturer in its motor racing endeavours, and has retained an almost permanent foothold in the sport, which it continues to this day with the Toyota Racing Series, and the Toyota 86 Series.

It's quite possible that almost every person in New Zealand holding a driver's licence has driven a Toyota at some stage, whether it be their own car, that of a friend or family member, a rental, or company car. They're hugely popular work-horses, from taxis to farm trucks. It's that dependability and quality that keeps bringing people back to Toyotas, time after time; whether they're a new car buyer, or a penniless student with an elderly weathered Corolla, and needing cheap and reliable transportation that'll get them to campus without incurring expensive maintenance bills.

In this modern motoring age, most car owners are not car enthusiasts. Indeed, to many, the motor vehicle is viewed simply as an appliance; a device for transporting its occupants from one point to another with the least amount of fuss. In this sense, almost every manufacturer on the planet is a producer of appliances. It's their bread and butter. But Toyota as a company is so much more. It's populated by genuine car enthusiasts, who have created some of the most truly impressive, most exhilarating vehicles in automotive history. And more so, many Toyota owners share that passion.

Indeed, this book is dedicated to Toyota enthusiasts, who have a genuine love for the brand. This is their book, and these are their stories.

Barry Crump

Barry Crump MBE was a writer and a poet. He penned more than 30 books, including *Wild Pork and Watercress*, which ultimately became the film *Hunt for the Wilderpeople*. And for any kid growing up in the 1980s and early '90s, he was the hardy, cigarette-toting bushman starring in the brilliant Toyota Hilux television commercials.

The Hilux adverts were a stroke of marketing genius. Significantly, when the first one aired in 1982, Toyota as a brand had been in New Zealand for barely 15 years. Still largely consuming a diet of big Australian sedans and British compacts, there was still a touch of distrust among Kiwi car buyers over Japanese products.

Crumpy's nuggety complexion, deep facial crevices, and raspy voice with that most twangy and Kiwi accent belied his deeper, more emotional side, that shone through in his writing and poetry. He was a living, breathing paradox.

There were four Hilux commercials starring Crump and Lloyd Scott (Scotty). The script for the first was fairly simple. Scotty, playing a Toyota agent, meets Crumpy on a dirt track to show off the shiny new third-generation Hilux, emphasising the 4WD option which Toyota had introduced to the model in 1979. Crump emerges from the bush, and is greeted by Scotty, who announced, 'Sorry I'm late Barry. Isn't she a beauty? Hilux four-wheel-drive.' Crump mosies down an embankment carrying a rifle, wearing a Swanndri and Akubra hat, and with a lit cigarette dangling out the corner of his mouth, and mumbles, 'Yeah it looks OK.'

From there, Crump slides his shotgun behind the seats, takes the wheel and proceeds to pound on the Hilux, scaling mountainsides and traversing rivers, and pressing ever deeper into the wilderness, while Scotty tries to impress by reciting all the features that make it so durable, emphasising the cross-technology with Toyota's rugged Land Cruiser. Every now and then his chauffeur will fire off a typical Crumpy one-liner. Finally, Scotty asks, 'How've you been getting in here before now?' To which comes the response, 'Well there is one other way you can get in here…', and as he slides to a halt alongside his grubby Land Cruiser, finishes with, 'and that's it right there.'

At just 60 seconds in length, the Hilux commercial was a total game-changer. It helped cement Toyota as a true-blue New Zealand product, built tough to suit the challenges of Kiwi living. But the winning try was the association with Barry Crump.

When the fourth-generation Hilux was launched in 1983, Toyota New Zealand quickly followed up with its next Crumpy and Scotty commercial. It opened with Scotty polishing a bright yellow Hilux and singing Toyota's theme song, before Crump came barrelling out of the bush in the same third-generation model that starred in the first commercial, skidding to a dusty halt.

Throwing his backpack in the rear tray of the new example, he looked in the window, plucked the lit cigarette from his mouth, and said, 'So this is it, eh?' Scotty, still polishing, proudly asked, 'Gidday, Barry, what do you think?', to which Crump responded, 'Bright, isn't it.' Before running up to the hut for a brew, Scotty grabbed a bag, disappeared behind a tree, and emerged wearing a shiny racing suit and helmet. 'Ahh, spare me days, mate, there's no need for all that, we've got a decent track here now.' The next 30 seconds cut from scene to scene as Crump put the new Hilux through its paces, with Scotty hanging on for dear life. The commercial ended with the mud-splattered Toyota skidding to a halt outside Crump's hut, cutting to Scotty, frozen with fear and half his face covered in mud. A voice-over finished with, 'The incredible new Toyota Hilux. It's so good, it's frightening.'

As with the first commercial, it hit the point home. The third commercial followed a quite different theme. It's often referred to as 'Scotty's Revenge'. Crump, looking entirely uncomfortable dressed in a suit and tie, arrives at Scotty's Toyota dealership in his muddy yellow Hilux, and called, 'Gidday, Scotty. Point me to St Paul's Church, mate, I'm a bit lost.' Scotty checked his watch, and responded with, 'Oh, come on, I'll take you.' As Crump climbed out of his Toyota, Scotty complimented him on his suit. 'Barry. Gee that's neat. New suit, eh?' Crump replied, 'They bury a mate of mine today.' A concerned Scotty queried if he was attending a funeral. 'No,' stated Crump, 'wedding!'

Scotty then ushers Crump into a new two-wheel-drive Hilux SR5 parked on his forecourt, before carrying out a spree of death-defying acts in amongst Wellington traffic. Yanked out of the bush, and out of his comfort zone, this time it was Crump who was holding on for dear life. Finally, the pair arrived at the church and, predictably, the groom-to-be runs outside, jumps in the back of the Hilux, and Crump proclaims, 'You'd better bury it, Scotty, looks like he's changed his mind again.' And with that, the three of them sped away.

Ultimately, much of Scotty's Revenge ended up on the cutting room floor, because the street scenes were considered to promote dangerous driving.

Crump and Scotty also teamed up in 1992 to sing the song 'Side By Side', written by Crump, and played as part of the Team New Zealand America's Cup yachting campaign, which Toyota New Zealand also helped sponsor.

Barry Crump was very much his own person, and certainly broke the mould. Having married five times, and spawned six sons, Crump died in 1996 at the age of 61.

Barry Crump and his wife Maggie.

Jodie and Drew Donovan's

Supras

Much like drag racing and American stock car racing, drifting is a sport whose underpinnings lay knee-deep in the highly illegal world of street racing. And likewise, although this gladiatorial contest, set inside a tight asphalt amphitheatre surrounded by cheering fans appears tailor-made to that of a spectator sport, it was never intended as such.

Drifting in its purest form dates back to the dawn of motor racing itself, but not as a sport in its own right. Rather, the limitations of skinny bias-ply tyres, and a complete absence of aerodynamic downforce, ensured race cars would tackle medium- and high-speed corners in a four-wheel drift as a matter of course. It wasn't until the early 1970s, as tyre and aerodynamic technology finally caught up with outright speed, that race cars would finally attack corners being pressed into the race track. And as a result, perhaps the most charismatic by-product of speed itself, the four-wheel drift, was consigned to history.

However, the sport of drifting as we know it, in which the central theme itself is based upon the driver's ability to hold their car in a near-constant slide while negotiating a set course, with points awarded based on the driver's ability to maintain speed and sideways balance, is believed to date back to 1970s Japan. What started life as an illegal underground contest was popularised in the 1980s by a racing driver named Keiichi Tsuchiya, the 'Drift King'.

But even before the Drift King rose to fame came another prominent Japanese racing driver by the name of Kunimitsu Takahashi who combined the mastery of drifting into his driving style, to better improve his lap times. Indirectly, and although he never took part in a drifting contest, Takahashi is today considered the 'father of drifting'.

Born in 1940, by the time he'd reached his early twenties, Kunimitsu Takahashi was carving out a highly successful motorcycle racing career. Aboard a 250cc Honda at the mighty

German race track Hockenheim, he became the first Japanese rider to win a world motorcycle Grand Prix, in 1961. But his career hit an abrupt end when he crashed heavily in the 1962 Isle of Man TT. Unable to continue racing motorcycles, he switched to four wheels in 1965, where he continued to succeed.

It was while racing a Nissan Skyline GT-R KPGC10 in the All Japan Touring Car Championship in the 1970s that Takahashi's driving style came to the attention of Japanese street racers. Rather than adhere to the traditional method of cornering, which involved attempting to maintain a balance of speed and grip, and without breaking rear traction, Takahashi instead dove deeply into the corners, braking later and letting the back of the car drift into a slide which he then held until well after he'd set off down the following straight. While the front tyres followed the racing line, the back tyres broke traction which was held in a controlled slide. This method, while unconventional, proved enormously successful for Takahashi, and he became a serial race winner. But it was also hugely entertaining to watch. His flamboyant driving style further cemented his hero status in Japan. And one of those who idolized him from the sidelines was a young Keiichi Tsuchiya, who began his own racing career in 1977.

Tsuchiya mastered his art as an illegal street racer. It was through local mountain roads (known as The Touge) in the dark of night that street racing took place in Japan, with most competitions being timed runs from one point to another. With courses lined with tight bends, racers began adopting the Takahashi style of drifting the car, Keiichi Tsuchiya among them. And it was Tsuchiya's ability to hold the car in an impressive slide while maintaining speed that earned him his reputation as the 'Dorikin' (Drift King).

Although he went on to become a successful professional racing driver, much as his hero Takahashi had done, Tsuchiya couldn't leave his illegal drifting roots behind completely, and in 1987 he created a ground-breaking short film called *Pluspy*. At just 23 minutes long, *Pluspy* starts at Mt Fuji racing circuit where Tsuchiya is racing his Toyota Corolla in the All Japan Touring Car Championship. It then leads into his other passion, and shows him climbing aboard his own modified Toyota AE86 Sprinter Trueno road car which he used for illegal drifting. The remainder of the film shows him drifting around various mountain roads.

Pluspy landed Tsuchiya in hot water with the authorities, who revoked his driver's licence. It briefly affected his racing career, although he quickly recovered and went on to enjoy ongoing success, competing for various manufacturers both in Japan and internationally, including at the Le Mans 24 Hours. He also became a consultant for the popular comic book series Initial D, based on Tsuchiya and his AE86 Sprinter.

The release of *Pluspy* suddenly thrust the art of drifting into the public spotlight, and very quickly it became the subject of numerous car magazines. It prompted Japanese tuning magazine *Carboy* to host the first ever legal drift competition shortly thereafter. Three years later, *Option* magazine, with guidance from Tsuchiya and the magazine's founder Daijiro Inada, created the first sanctioned drift competition, called Ikaten, in which teams of five cars drove in synchronized drifts to entertain spectators.

The Ikaten events increased in popularity, and in 2000, Inada established the All Japan Professional Drift Championship. The first event took place at Ebisu Circuit in Fukushima, and had over 40 contestants who all made individual runs on which they were judged. This event then counted as the first round of the 2001 Championship which became a multi-round competition. From Round 2 onwards, the format changed to a head-to-head combat in which two cars took part in two passes, with the judged winner moving on to the next round, and the loser being eliminated from the event. This format is still used in all drift competitions throughout the world. The All Japan Professional Drift Championship was renamed D1 Grand Prix.

Unlike traditional forms of motorsport where an assembly of cars races over a set distance or number of laps, with the first car to complete the distance being declared the winner, drifting competitions are not so clear-cut. Much like freestyle motocross, drifting is judged based on vehicle control and flamboyance. Each contest has two cars set off together, one in front of the other.

Predictably, with the Japanese performance car scene becoming so firmly entrenched in Kiwi car culture over the past few decades, the sport of drifting was eventually adopted down-under. Much like the US and Australia, where drifting has also gained a huge following, not all the cars are Japanese. Indeed, in the US, Mustangs, Camaros, and even Corvettes are common. The cars need to be rear-wheel-drive, and because not everyone wants to drive a Nissan, the NZ drifting scene has a handful of American and Australian muscle cars competing. But by and large, Nissan are the most

popular choice. Fittingly, however, given that Keiichi Tsuchiya and his little AE86 played such a pivotal role in giving drifting the widespread recognition it deserves, Toyota has a strong presence in the New Zealand drifting scene too, thanks to the pair of insane 1000-horsepower Supras of husband and wife team Jodie and Drew Donovan, of Team DMNZ.

Drew started drifting in 2007, Jodie a year later. 'The motion of sliding a car is pretty enjoyable to say the least', explains Drew. 'And I think as long as there have been rear-wheel-drive cars there has been drifting of some sort. But for me, the first time I saw drifting was on late night TV, I couldn't believe it was a sport, and I wanted in. I first hit the racetrack to try drifting in 2007. And with Jodie being a bit of a petrolhead, it didn't take much convincing to get her behind the wheel in 2008, and she got hooked on drifting as well.'

Unlike most forms of motorsport, which even novice race-goers can quickly grasp, drifting is far more complex. The winner is rarely clear-cut, unless you really know what you're watching. Unless you follow drifting, you're likely not familiar with how it works, and how a winner is decided upon. Jarod Deanda of Formula Drift in the United States probably explains it best: 'Drifting is the only motor sport in the world that is based on judging. In head-to-head battles, the lead driver is supposed to run a specific judging line, with as much angle as possible. The lead driver is setting the pace, but the chase driver is supposed to utilize the lead driver as a moving clipping point; where the lead driver goes, the chase driver should go too. After two runs with each driver leading once, the driver that was the most consistent, the most aggressive, and was able to show clear dominance over the other driver, will win the match-up.'

To further complicate matters, the lead driver must remain consistent; they can't just speed up and slow down, because the chase driver needs to follow them as closely as possible, without actually hitting them, and preferably, hold their car in a more aggressive drift. The lead driver is attempting to negotiate the course as quickly as they can, but with the tail of the car at as much of an angle as they are able, and with the rear tyres producing as much smoke as possible. If the lead driver can establish a sizeable gap to the chase driver, but while still keeping their car in an aggressive drift, this is considered favourable. The lead driver can't just trundle, even if they've done a good job of stepping the rear of their car out. Drifting is a combination of speed and style. Likewise, if the lead driver is displaying less angle than the chase driver, this is deemed favourable for the chase driver.

If the chase driver hits the lead driver, the chase driver is judged to have lost. However, if the chase driver hits the lead driver because the lead driver slowed suddenly, then blame is apportioned to the lead driver. If either driver loses control and spins or crashes, they'll be judged to have lost. If after the two runs the judges can't make a clear decision on who performed best, they'll do another two runs.

Drifting may have started on Japanese mountain roads by street racers in their modified road cars, but modern-day drift cars are full-blown, purpose-built race cars, designed specifically for drift competitions. The cars themselves look amazing, and hugely aggressive. Although most start life as production cars, not a lot of the original production car is actually retained in the finished race car. The interiors are completely gutted, and fitted with a rigid rollcage. The cars sit very low to the ground, and for the most part, roll on big-diameter wheels and low-profile tyres. The bodywork is usually widened with the fitment of wide wheel-arch flares, and clip-on deep front clips, and large rear wings are also attached. The entire body is then usually cloaked in a vivid vinyl wrap, displaying the team's various sponsors and graphics.

What's not so obvious from the outside are the various specific requirements built into a drift car in order for it to be successful, such as the super-fast steering system which allows the driver to make quick steering adjustments, as well as the heavy front negative camber. Drifting contests invariably use track layouts that have constant direction changes, to ensure the cars stay in a constant slide, and the driver needs to make those directional changes as quickly as possible. In addition, the steering allows a much greater degree of front-wheel angle, so the car is positioned in a more aggressive slide. Also, the driver has a hand-operated brake, which briefly locks the rear wheels to help speed up directional changes, and control the angle of the slide. These are hugely technical cars.

Many drift cars have upwards of 1000 horsepower. This allows them to get quickly up to speed from the standing start, and then easily break the rear tyres into a wheelspin and slide, and be able to keep them spinning. Drift cars will take faster corners in excess of 100 km/h. Remember, that's with the car held in a full tyre smoking slide, with the back-end swung right out. Drifters are highly skilled. Although there is no hard and fast rule, it seems most drifters prefer

a car with a wheelbase anywhere between 98 inches (2489mm) and 108 inches (2743mm). If the wheelbase is too short, the car becomes too twitchy and too difficult to hold in a consistent drift. Too long and it becomes cumbersome. Most drift runs only last around 30 seconds, by which time a new pair of rear tyres can be worn down to the canvas. It's hugely impressive, and takes gobs of talent. Crashes, spins, and body contact are common. The cars look cool, but they take a beating.

Drew's first drift car, albeit unofficially, was a 1980 Toyota Corona, which he got at age 11. Since then he has owned everything from Ford Falcons to Mazda RX7s, and plenty in between. But he keeps coming back to his beloved Toyotas. Both he and Jodie drive JZA80 Supras. Why did they choose Supras? 'Why wouldn't you?' says Drew. 'Honestly, they are the best chassis I've drifted and have so much potential for loads of rear grip. We did the bulk of the build work ourselves but with plenty of help from our crew and family.' Both Supras started life as road-going production cars, but have been massively modified for drifting. 'Between the front and rear strut towers the factory floor remains, bar the driveshaft tunnel. I'd say 70% of the floor/unibody chassis, roof and doors remains. The rest is fibreglass and carbon. Front and rear [beyond the front and rear shock towers] tube work is essential, and is a lot easier to repair.'

Drew and Jodie's Supras may appear similar at first glance, bar the colour changes in the body graphics, but there is a variety of differences. Jodie's (nicknamed Bonnie) has a BRZ86 front clip attached to give her car a more contemporary appearance. Bonnie also has a rear boot-mounted spoiler, whereas Drew's (nicknamed Clyde) retains an aftermarket version of the standard nose, and a tall rear wing. The changes beneath the skin are even more dramatic. Bonnie is fitted with a stroked 3400cc (207 cu in) 2JZ GTE Toyota inline-6 engine fitted with a Garrett GT42 turbocharger, to give 1000 horsepower. Clyde, by contrast, runs a 7400cc (454 cu in) Chevy LSX V8, with Whipple 2900 supercharger, and producing a whopping 1090 horsepower. Why is it so important to have such big horsepower in a drift car? 'A good drift car is the perfect balance of grip, weight and power,' explains Drew. 'The more power you have, the more grip you can tune in, and the faster you can run.'

Both cars use NASCAR-type gearboxes, Hakon Racing suspension, Winters Performance quick-change differentials, 18 x 8 and 18 x 11 inch TBC Performance wheels, and modified Wisefab steering kits. Likewise, both weigh around 1180kg. '1180 is just as light as we could get the cars on our budget,' explains Drew. 'New Zealand doesn't have weight rules like the US, yet.' Drifting regulations require cars be fitted with DOT rated tyres, so Bonnie and Clyde are fitted with Zestino Gredge 07RSs. 'We have to run a DOT rated tyre no larger than 265. We run a 265-35-18 semi-slick race compound tyre for maximum grip'. And how many sets of tyres do Drew and Jodie usually go through in an event? 'We take 60 tyres away with us, and on a good weekend we will use them all.' What have been Team DMNZ's best results to date? 'I've had a couple of first place podiums and plenty of other lower down, but the D1NZ Championship trophy is what I'm really after.' And what has been the personal highlight to date? 'We have so much fun doing what we do, but I think winning the home round of D1NZ at Bay Park in 2018 would have to be up there.' And what about future goals? 'The D1NZ Championship and more driving overseas is the focus right now. Also, I'd love to pair up with Jodie in an enduro team or something similar. But our sport takes every dollar I can find, so it couldn't be my own car. But I'd be in quicker than a rat up a drainpipe if the offer came.'

Drew and Jodie Donovan's monster Supras are about as different from Keiichi Tsuchiya's modest little AE86 as it's possible to be, such is the level at which drifting has developed over the past three decades. But it's good to see this committed and talented husband and wife duo carrying the Toyota name with pride in New Zealand. The 'Drift King' would be proud.

D1NZ
Husqvarna
D1NZ
haxon
Get the Juice!
YOUHOST
TBC
RED STAR SIGNS

The Little Red Corona

This modest little RT40 Corona, registration DY 5996, is perhaps the most significant Toyota in New Zealand. This is believed to be the oldest existing New Zealand assembled Toyota; one of the very first to roll off a New Zealand assembly line, if not *the* first.

DY 5996 came off the Steel Motor Assembly Plant in Christchurch in early 1967, and was first registered on 3 February. Its first owner was Raymond James Dalton, who kept it until 1971 when it went to Waikato Motors. The next owners, a young married couple, kept the car until 1976, when Toyota tracked it down in a search to find its oldest surviving car. Toyota offered the couple a brand new RT102 Corona in exchange for the nearly-ten-year-old RT40, which had clocked up a relatively high mileage, and was well-used. DY 5996 was then sent to Toyota's SKD assembly plant in Johnsonville, Wellington, to be restored.

Typical of the early New Zealand Toyotas, the Corona is a case of function over form. Its 1490cc 4-cylinder engine produces 82 horsepower (61kW), which is backed by a 4-speed manual transmission, and rides on 13-inch-diameter wheels, while stopping power comes from drum brakes in all four corners. Although the Corona was Toyota's largest offering in 1967, by today's standards, it's tiny, even compared to a modern Corolla. Its styling is incredibly unassuming. Toyota was selective in its product placement as it sought to slowly infiltrate this new market made up of buyers who had traditionally driven British, European, and American cars.

Toyota owns the RT40 to this day, and has it displayed at the company's head office. It's road registered, driven infrequently, and makes occasional appearances at car shows, including the Toyota Festival.

DY 5996 may not have the physical presence of a Lexus LC 500, but for Toyota in New Zealand, this is where it all began, and this car generates crowds wherever it goes.

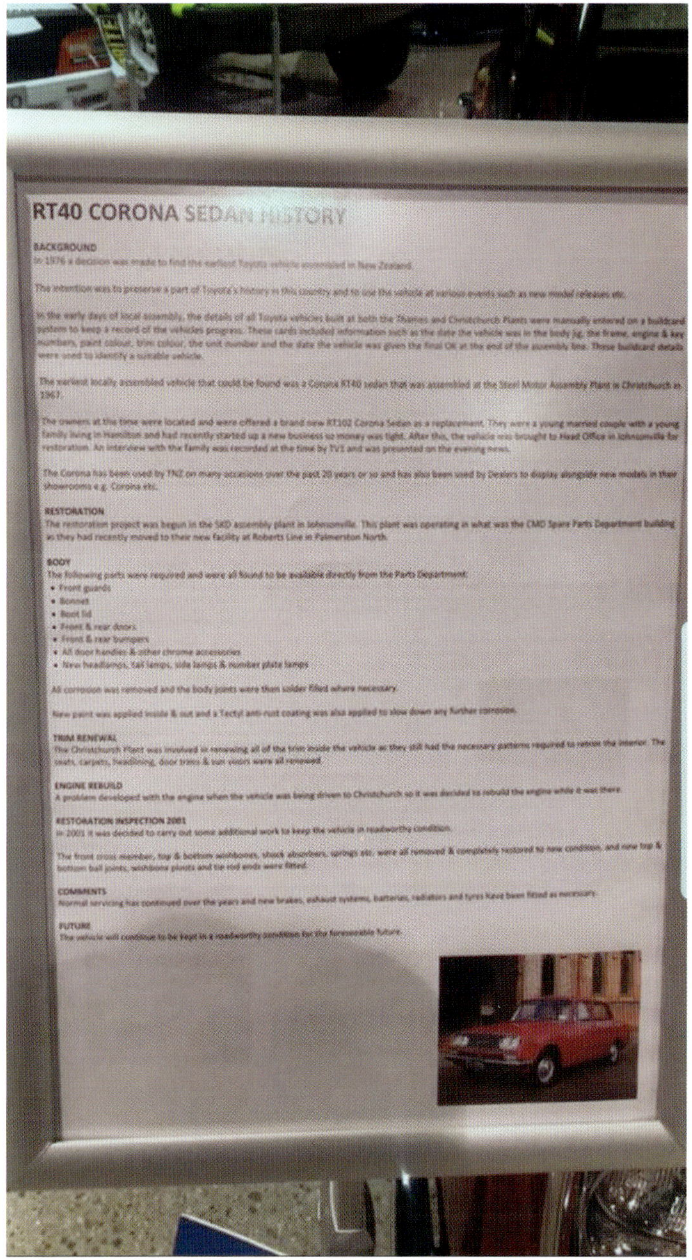

Corona
TOYOTA
DY 5996

A modest little car it may be, but its significance in Kiwi motoring history greatly exceeds its diminutive proportions.

McLaren Toyota

During the late 1960s and early 1970s, the Sports Car Club of America (SCCA) organized the Canadian American Challenge Cup (Can-Am) for Federation Internationale de l'Automobile Group 7 unlimited sports cars. It was one of the most successful, most spectacular motor racing series on the planet. And one of the richest too. Indeed, inaugural Can-Am champion John Surtees collected around US$78,000 prize money in 1966, significantly more than if he'd managed to win every race in that year's Formula 1 World Championship. So naturally, the Can-Am held a strong magnetic force, attracting many of the world's best drivers and teams, as well as car manufacturers. And those elements, understandably, attracted race fans and the media.

The Can-Am series boasted a world class driver entry, which over the years included top drivers from around the world. But leading them all, for the most part, was a band of Kiwis, headed by the brilliant New Zealand driver/engineer, Bruce McLaren.

Can-Am series regulations were virtually limitless. In order that they be identifiable as sports cars, there were minimum cockpit and door size requirements, and there had to be two seats positioned side by side. Otherwise, designers were given total creative freedom to build the fastest, most powerful racing cars on earth.

Such was the open-formula approach to Can-Am, it was a world leader in engine, chassis and aerodynamic development. Turbo-charging, ground-effects, and moving aero devices all appeared in Can-Am long before they did in Formula 1. Likewise, for much of its history, Can-Am cars were faster than Formula 1 cars on many of the tracks the two divisions visited.

For the young Kiwi, Bruce McLaren, the Can-Am proved instrumental in the growth, funding, and reputation of the racing team which he'd established just a few years prior. McLaren and Chris Amon drove factory McLaren cars in the 1966 Can-Am, before Amon took up an offer to drive for Ferrari. His place was taken by another Kiwi, Denny Hulme. Neither McLaren nor Amon won a race in 1966. However, the new-for-1967 McLaren, the M6A, proved to be a highly effective piece of kit. Of the six races, Hulme won three, McLaren two, and McLaren took the title ahead of his team-mate. This set the scene for the next four years.

In 1968, the new McLaren M8A was comfortably the fastest car on the grid. Hulme won the championship, with three victories, while McLaren finished second, with one. Only mechanical dramas prevented a total whitewash. But more was to come. 1969 saw the Can-Am series extended to 11 races, and the McLaren team, with its impressive M8B model, won all of them. Bruce McLaren won the championship, with six victories, while Hulme took five wins to finish second. By the end of the 1969 season, the dominance of the McLaren cars saw the Can-Am series affectionately dubbed 'The Bruce and Denny Show'.

Bruce McLaren was tragically killed in a testing accident at Goodwood in June 1970, but, such had been the success of his team that it was able to continue without its founder at the helm. Denny Hulme won the 1970 Can-Am series, while in 1971, new McLaren recruit Peter Revson took McLaren's final Can-Am Championship.

The McLaren team in the Can-Am drew lucrative sponsorship deals, but also won large prize purses at each event. In addition, a business relationship was struck with Peter Agg, through his Trojan company in England. Trojan manufactured tiny micro-cars for the British and European markets, as well as Elva sports cars. The McLaren/Trojan partnership led to replicas of the factory McLarens

Hiroshi Fushida poses with the McLaren-Toyota he was to drive at the 1969 World Challenge Cup Fuji 200. Fushida supplied several of the period images of this car to its owner/restorer, Duncan Fox.

built by Trojan for sale to customers, much as rivals Lola were doing. Due to this successful partnership, most grids throughout the history of the Can-Am series were largely made up of McLaren customer cars.

In truth, the majority of Can-Am races were processional and largely predictable competitions due to McLaren's dominance through to 1971, followed by that of Porsche in 1972 and '73. But somehow, the lack of close-quarters racing was forgiven. For while the star-cast of drivers was an important ingredient in its success, the cars themselves were just as much a factor with Chaparral, Lola, Ferrari, Autocoast and Shadow challenging for victory. Can-Am cars were intoxicating, addictive even, making this one of the most successful racing categories in history. And because of this, attempts were made to replicate the success of the Can-Am series in other parts of the world, including Europe and Asia.

In 1968, Group 7 was introduced to Japan and, although much less known internationally than the North American Can-Am series, Japan Group 7 enjoyed fierce competition, and intense factory participation from Nissan and Toyota/Yamaha. Even Isuzu joined the party. There were a number of high-profile sports racing events contested in the late 1960s in Japan for Group 7 cars, most notably the Japan Grand Prix and World Challenge Cup Fuji 200, both held at the perilously fast Mt Fuji Speedway.

After decades of manufacturing mostly utilitarian vehicles and diminutive city cars, Japanese car makers began providing more upmarket offerings by the early 1960s, as the economy soared. In 1963, the very first international racing event contested in Japan was held at Honda Motor Company's test track, Suzuka. This event, held for production sports and touring cars, was named the Japan Grand Prix. Its success helped ignite massive enthusiasm in Japan for motor racing and sporty performance vehicles. Although dominated by European teams and manufacturers such as Ferrari, Lotus, Aston Martin, and Jaguar, there was a healthy Japanese presence in the inaugural Japan Grand Prix, most notably, the Skyline GT, built by the Prince Motor Company.

There was no Japan Grand Prix in 1965, and in 1966, the event moved to the newly constructed Mt Fuji Speedway, which quickly became Japan's premier racing facility. Having been defeated by Soukichi's Porsche 904 in 1964, Prince Motor Company sought to build a similar mid-engined car, based on a Brabham chassis, and powered by the 2-litre inline-6 Skyline motor. Named the R380, it went on to beat Porsche in the 1966 Japan Grand Prix, and underwent further development when Prince Motor Company was merged into Nissan that same year.

But while Nissan/Prince were taking on the European invaders, and gaining huge national exposure in Japan as a result, Toyota was virtually anonymous in both motor racing and the new sporty car market. The tiny 2-cylinder Sports 800 was released in 1965 as an entrée, but Toyota finally began getting serious with the release of its beautiful 6-cylinder front-engined 2000GT in 1967. The 2000GT enjoyed success in endurance racing, winning the 1967 Fuji 24 Hours as well as the Suzuka 1000km but it wasn't competitive enough in the all-important Japan Grand Prix.

Motorsport in Japan was run by the Japanese Automobile Federation (JAF), which was recognized by the FIA as domestic racing competition began attracting international interest. FIA Group 7 unlimited sports car racing regulations were introduced in Japan for 1968. And this would prompt a full-scale war between Toyota/Yamaha and Nissan as Japan's leader in domestic motorsport.

Toyota were on the back-foot, with Nissan already stealing the jump on them. They partnered with Yamaha to build the company's first Group 7 car, the 415S, designed by Jiro Kawano. Nissan had continued development of the Prince-built R380 and for 1968, it unleashed the impressive R381, built for the first time to Group 7 regulations. Although Nissan planned to use a new Prince Motor Company developed 6-litre V12 (essentially two inline-6 Nissan engines joined together), delays in its completion instead saw the fleet of Nissan R381s fitted with small block Chevrolet V8s. Their most prominent feature was a pair of tall rear flipper wings mounted side by side, and actuated by hydraulics which tilted their angle based on suspension load.

By comparison, the Toyota 415S was small, modest, and, alongside its Nissan rival, slightly underwhelming. Its jewel was a magnificent new 3-litre quad-cam Toyota V8, designed and manufactured by Yamaha, but with Toyota badging, and using an aluminium block, flat-plane crankshaft, and fuel-injection. But at around 300 horsepower, it was at least 150 horsepower shy of the Chevy V8 in the rear of the Nissans. Despite its small displacement, the Toyota V8, with its 90-degree 'V', was around 160mm wider than a big block Chevrolet. The chassis featured an aluminium tubular space-frame, and bodywork which bore a number of similarities to the Lola T70 which had won the inaugural Can-Am Championship.

The little Toyota 415S was a multiple race winner in 1968, in the lesser supported events. For Nissan, the only race that mattered

Pictured at the 1969 World Challenge Cup Fuji 200 is the McLaren Toyota M12. This was its only racing event under Toyota's ownership.

was the Japan Grand Prix. This event carried more prestige than the rest of the year combined. And naturally, come the Japan Grand Prix, Nissan destroyed its rival. Two of the Chevrolet-powered R381s were entered, and they qualified first and second. One of the Nissans suffered delays during the race, but the other won handily, from Tetsu Ikuzawa's Porsche 910. The first Toyota home was down in eighth, and several laps behind the winner.

In October 1968, a new event was launched, called the World Challenge Cup Fuji 200. This was an epic 200-mile (320km) contest in which several top cars from the Can-Am series were invited to take part. There were no factory McLarens, but Mark Donohue, Peter Revson, Jerry Titus, and Jo Bonnier were all aboard privateer McLaren M6s. Numerous upgrades had been made to the Toyotas since the Grand Prix in May, but even so the monstrously powerful 7-litre big block Can-Am cars virtually blew them right off the race track.

Five Toyota 415Ss were entered, and impressively, all five reached the finish. Nobody could question Toyota's impeccable reliability. They were almost mechanically indestructible. But they just weren't fast enough. The best Toyota finisher was Yukio Fukuzawa, in fourth. But despite a trouble-free race, he was lapped seven times by race winner Revson, in a 75-lap race.

Radical changes were needed for 1969. But Toyota were nothing if not dedicated to achieving success. There were now more than 100 Toyota and Yamaha staff working on the racing programmes, and 1969 would see huge strides made in every area.

1969 did not start well when Toyota's star driver Yukio Fukuzawa was killed while testing the 400S Group 6 prototype coupe at Yamaha's test track Fukuroi. The 400S, originally labelled the JP6, was the creation of Pete Brock, who, at the age of just 19, designed for General Motors what would become the 1963 Corvette Sting Ray. He went on to transform the Shelby Cobra roadster into the successful Shelby Daytona Coupe, while working for Shelby American in 1964.

Brock established Brock Racing Enterprises (BRE) in late 1965, and by 1967, was doing work for Japanese truck makers, Hino. Hino had recently branched out into the car market, and wanted to use motor racing as part of its marketing. To that end, the company enlisted the help of BRE, kicking off with the tiny Contessa, which took a surprising victory in the 1966 Mission Bell 100, a support race at the LA Times Grand Prix event, the Riverside round of the Can-Am series championship. Brock then created the futuristic 1.3-litre BRE Hino Samurai GT sports racing car. The Samurai featured a movable rear wing, 20 years before Ferrari sported a similar wing on the back of its F40. When Hino was taken over by Toyota in late 1967 following the death of its president, Toyota offered Brock the opportunity to create an FIA Group 6 endurance racing sports car, powered by its 3-litre quad-cam V8. Brock designed the visionary Toyota JP6, which incorporated highly advanced aerodynamic and chassis characteristics. The body featured movable aerodynamic devices that pressed down onto the chassis as speed increased, thus providing greater adhesion the faster the car went.

However, in a surprise move, Carroll Shelby struck up an agreement with Toyota to race its 2000GT in the US in 1968, the flow-on effect being that Toyota promptly ended its relationship with BRE. The partially completed JP6 was seized and sent to Japan where Toyota's engineers completed its construction. Brock was concerned that Toyota engineers might not understand the car's highly complex design and the relationship between body and chassis. It crashed during its first test. The implications of Fukuzawa's fatal accident would later affect Toyota's involvement in Group 7.

Toyota's Group 7 weapon for the 1969 season was the new 474S. Jiro Kawano's latest creation was longer and wider, with a more rigid tube-frame chassis than the 415S. It also featured a contemporary wedge-shaped body, as seen on the latest Can-Am cars. The nose was long and sharp, allowing air to travel smoothly over it to the rear spoiler, while the sides were flat and deep. Importantly, the little 3-litre quad-cam Toyota V8 was enlarged to 5 litres, and now produced around 550 horsepower at 8000rpm. It was backed by a 5-speed Hewland transaxle. Another home-run for Toyota was in securing the services of highly accomplished racing driver Vic Elford, who would help with the new car's development. Elford had experience in numerous European and American racing categories, and he'd been a factory driver for Porsche. Toyota needed all the help it could get; chief rivals Nissan were now ready with their latest Group 7 machine, the incredible 6-litre V12-powered R382.

As usual, Nissan opted to skip most events in 1969, and in most cases, Toyota claimed hollow victories. The race that mattered was the 1969 Japan Grand Prix, which this year had been moved to October, in a bid to entice more international entries.

Nissan entered three R382s as part of the 32-car grid, while Toyota had a quartet of 474Ss. Up against the local manufacturers at the Japan Grand Prix was Porsche. The Stuttgart firm sent one of its latest factory machines, the 917, to be driven by Jo Siffert and David Piper. In qualifying, Nissan dominated, with its three cars claiming the top three start positions. The fastest Toyota entry, in fourth, was that of Hiroyuki Kukidome/Shihomi Hosoya, but this pairing was nearly 3.5 seconds off pole. Two more Toyotas followed, with the Porsche 917 next, ahead of the fourth Toyota.

Come the race, which was just over 500 kilometres in length, and Toyota flew out of the blocks. From the standing start, there were three Toyotas in the top four as the field streamed into the high-speed banked first turn, split only by the Porsche. But that was as good as it got. As the race progressed, so Nissan asserted its dominance, and surged forward. By the finish it was Nissan in first and second, with the best of the Toyotas, that driven by Hiroshi Fushida/Minoru Kawai, in third, one lap behind the winners. Toyota had suffered defeat once more. But the gap had clearly reduced.

The following month, in November 1969, Toyota entered the World Challenge Cup Fuji 200. Rivals Nissan were absent, but this didn't matter, as Toyota could now test its new machinery against some of the fastest racing cars on the planet, and evaluate just how much progress had been made. Moreover, Jiro Kawano had completely re-designed the bodywork on the fleet of Toyota 474Ss. The 1969 Japan Grand Prix had shown the new Toyotas to produce too much drag for the fast Mt Fuji Speedway layout. Mt Fuji was a monster of a race track originally conceived as a 4-kilometre-long NASCAR-style banked oval super-speedway. However, there was a funding shortfall following completion of the first banked turn, after which the rest of the layout resumed that of a more traditional racing circuit with various lightly cambered flowing left and right turns linked by short and medium-length straights. The final turn, although only lightly cambered, was incredibly fast, and led back onto the epic 1.5-kilometre-long Start/Finish straight which then plunged full-tilt into the high-speed 30-degree banked first corner once more. The fastest cars were entering the banking at 300 km/h.

Wisely, the World Challenge Cup Fuji 200 in both 1968 and 1969 was run in the reverse direction, to avoid the Can-Am cars plunging into the fast turn one banking at breakneck speed. But regardless, Mt Fuji Speedway was one of the fastest race tracks on the planet. It had already been the scene of multiple racing fatalities, but as Japan's most prominent venue, all the important competitions were held here. And it was imperative that designers build high-speed, low-drag cars to suit.

But while Nissan had achieved a low-drag design for 1969, Toyota, evidently, had not. The 474S featured bodywork that wrapped around behind the rear wheels, with a flat sectioned panel spanning the full height and width of the car, adjoining the tall rear spoiler. The aggressively angled spoiler provided good rear downforce, but it also produced a lot of drag, which limited top speed. At Mt Fuji Speedway, this was not a good thing. However, come the 1969 World Challenge Cup Fuji 200, the Toyotas looked vastly different from a month earlier. The blunt tail section had essentially been lopped off at the rearward point of the rear tyres, creating an open Manx-tail layout. Furthermore, the tall upright rear spoiler was also gone, now replaced by an adjustable aerofoil mounted on struts, set in a low-downforce position. In addition, more power was being coaxed from the little 5-litre V8. Drivers now had 600 horsepower at their disposal, although this was still at least 100 horsepower less than the big block Can-Am visitors.

Toyota entered four cars in the World Challenge Cup Fuji 200. Three of these were the updated 474Ss. The other was something of a surprise. Although adorned in Toyota racing paint, wearing giant Toyota lettering on its rear wing, a small Yamaha decal on its flanks, and fitted with a Toyota quad-cam 5-litre V8, the fourth Toyota entry was a McLaren M12.

The M12 was the McLaren/Trojan customer model for 1969. It featured a monocoque chassis based largely on the 1967 M6A. The M12, however, featured unique bodywork which closely resembled that of the road-going M6 GT Coupe Bruce McLaren had envisaged as part of his fledgling company's future growth. But unlike the M6 coupe, the M12 Group 7 racer's bodywork was in topless 'Spyda' form. Trojan supplied the cars as a rolling chassis, but set up for a big block Chevrolet, which the customer would source and fit themselves. Fewer than a dozen McLaren M12 chassis tubs were actually produced (including two which were designated for Bruce McLaren's M6 GT project), from which around nine became complete rolling chassis. Of these, two rollers were sold to Le Mans Co, the McLaren agent in Japan. One of them was purchased by Toyota Motor Company, seemingly as part of the company's R&D programme to help fast-track its own Group 7 aspirations.

Interestingly, the lone Toyota-entered McLaren M12 featured the same rear bodywork treatment as the in-house 474Ss, which differed significantly from the standard M12 bodywork. The enormous rear wing was attached to a frame on the rear of the transaxle, and measured over two metres wide. It was set at a very flat, low-drag angle. Clearly, Jiro Kawano was backing his new low-drag approach. And he was about to be proven right.

As expected, one of the Can-Am visitors, the Ti22 of Jackie Oliver, was fastest in qualifying. But what wasn't expected was that the McLaren/Toyota M12, driven by Hiroshi Fushida, was second fastest, and just two-tenths of a second behind Oliver's 7-litre big block monster. Minoru Kawai was third in the best of the 474Ss, followed by Revson, Cannon, and Parsons. This was an impressive performance by the Toyotas, and really underlined the enormous progress Toyota/Yamaha had made in the past 12 months. The race itself was to prove even more fruitful. Kawai won from Cannon and Motschenbacher. Toyota had good reason to celebrate.

Sure, Toyota was racing on home turf, with cars specifically designed for the unique Fuji layout and Nissan was absent. But just 12 months earlier, the best-placed Toyota was a full seven laps behind the race winner at the end. Now the powerful American Can-Am machines were giving chase to the Manx-tailed Toyotas. This was really the first significant international race win in the history of Toyota Motor Company.

Moments after the start of the 1969 World Challenge Cup Fuji 200, the McLaren Toyota M12 of Hiroshi Fushida (left of shot) jumps out to a small lead from the pole-sitting Ti22 of Jackie Oliver (#22), the Lola T163 of Peter Revson (#31), and the Toyota 474S of Minoru Kawai (#8).

This success provided confidence to tackle the next chapter in Toyota's ambitious plans; the Can-Am cars in their own backyard. To compete in the highly competitive Can-Am series, and better still, to actually win races, would be the next important step in infiltrating the lucrative American automotive market.

Toyota began selling cars into the United States in 1958. The first model that went on sale, the curious Toyopet Crown, found only 290 buyers, total. A decade later, the US market was still proving a hard nut to crack. When Ford, General Motors, Chrysler, and American Motors were embroiled in a horsepower race producing big, loud, powerful muscle cars, Toyota's most lavish model, the dinky Corona, couldn't have been more at odds. But through motor racing, and the loud, colourful, brash Can-Am series, Toyota could showcase its brand as it sought to make its products more widely accepted by an American audience.

But merely keeping company with the numerous McLaren/Trojan and Lola customer cars in the middle of the pack wasn't part of Toyota's plan. Despite having only built its first Group 7 car less than two years earlier, it now wanted to beat the best Can-Am heavy hitters. The problem, however, was the Toyota V8 engine. With an aluminium block, quad-cam cylinder heads, flat-plane crankshaft and fuel-injection, it was a wondrous piece of technology. But it wasn't nearly powerful enough to tackle the massive 7-litre big block units being produced by Detroit. The Toyota V8, in its current form, was nearing its potential limit, and stretching its capacity wasn't an option. To that end, Toyota turned to the highly complex and, in 1969, virtually unknown science of turbocharging.

So complex was the art of turbocharging in 1969, that few actually even attempted it. It really required the resources of a car manufacturer, who could design the engine from scratch. Simply bolting turbochargers to an existing engine wouldn't work. Toyota felt it had the ability to do this, and began a programme to turbocharge its quad-cam 5-litre V8. For this engine, a new car was designed, named the 578A. Again, it continued the wedge-shape that had worked so well on the revamped 474S, and was designed from the outset with a Manx-tail and adjustable rear aerofoil. To cope with the extra power and torque that would come from the turbocharged V8, the Hewland transaxle was replaced by a heavily reinforced Aisin Seiki unit.

The twin-turbocharged (using a pair of Garrett AiResearch turbines) Toyota V8 was nothing short of a masterpiece. At 8000rpm, it produced 800 horsepower, and 534 lb ft of torque. This was perhaps the most powerful circuit racing engine in the world at the time. And there was potential for more power. But taming the beast would really test the Toyota/Yamaha engineers. Much of the first half of 1970 was spent testing the 578A. It weighed around 650 kilograms, and with 800 horsepower, possessed an immense power-to-weight ratio. But understanding the turbo system, the cumbersome turbo-lag, and the sudden delivery of furious power when the turbos kicked in, was enormously challenging. The drivers were grappling with a wild animal.

It was while testing the turbocharged 578A at Suzuka in August 1970 that star Toyota driver Minoru Kawai lost control and crashed. He succumbed to his injuries.

With a second of its drivers having lost his life in less than 12 months, Toyota cancelled the 578A programme, and its plans to enter the Can-Am series. Its decision was made easier thanks to regulation changes made to the Japan Grand Prix for 1970. This became a race for single-seaters, as part of a longer-term plan to host a round of the FIA World Formula 1 Championship, which it ultimately succeeded in achieving, from 1976. Because of this, Nissan no longer sought to compete. Toyota's great racing rival was gone, and so had Toyota's motivation to compete. The 578A never raced, and Toyota would not produce another sports prototype racing car until the 1980s.

In 1972, the combined might of Porsche and Penske Racing entered the Can-Am series with a twin-turbocharged 12-cylinder Porsche 917/10, and proceeded to blow the previously dominant McLaren team right off the track. Porsche is widely considered to be the first manufacturer to finally unlock the massive potential of turbocharging in circuit racing. But Porsche were two years behind Toyota, and although the world will never know if Toyota could have fully tamed the art of turbocharging and gone on to dominate the Can-Am series as Porsche would later do, given the massive technological progress the company had made in the two short years from 1968 to 1970, it seems almost inevitable it would have.

Fushida battles the Lola T163 Can-Am car of visiting American Gary Wilson.

With sports cars no longer eligible for the Japan Grand Prix, the various Toyota Group 7 machines were ultimately destined to become museum pieces. But not so the McLaren/Toyota M12 that was such a sensation in the 1969 World Challenge Cup Fuji 200. This appears to be the only time Toyota raced the car. Following its impressive qualifying performance, the McLaren/Toyota retired during the race from overheating. After being used for further testing, during which it was retro-fitted with its original M12 rear bodywork, but with another wing attached, it was eventually placed in storage, without its engine.

Meanwhile, Group 7 racing continued in Japan, but without manufacturer participation, it was now a category of lesser significance. Of those competing in Japan Group 7 was a team entered by Kurasawa steel mill, who ran a pair of Chevrolet big block powered McLaren M12s. One of these was a car leased from Lothar Motschenbacher. The other was the second of the two original M12s imported by Le Mans Co in 1969. The team's driver was Tadashi Sakai. When the mill ran into financial troubles, it ceased its racing operation, at which point Sakai purchased the team. He then promptly crashed the ex-Kurasawa car. Through contacts at Toyota, he was able to purchase the McLaren/Toyota M12 tub and other parts, into which he fitted the running gear from the wrecked Kurasawa M12.

Sakai ran a two-car team of McLaren M12s through to the end of 1973 and enjoyed multiple race victories, until the car was retired from active duties. Upon its retirement, the ex-Toyota M12 ended up being housed in a giant glass case in a restaurant in Tokyo, where it sat for the next 25 years. The restaurant owner was close friends with the family of Japanese racing driver Hiroshi Kazato, who was killed at Mt Fuji Speedway in 1974 during a Fuji Grand Champion race when he collided with Seiichi Suzuki on the infamous turn one high-speed banking.

Even though Kazato is believed never to have driven the M12, the restaurant owner purchased it, along with various other racing cars, and created a shrine to celebrate the career of his friend where it was encased and on display for visitors to the restaurant. And it was there that the McLaren sat until 1999, when it was purchased by New Zealanders Duncan Fox and Tony Roberts. 'We heard about it through a Japanese doctor that we know,' recalls Duncan Fox. 'He had an interest in motorsport, said he knew the car, knew the owner, and were we interested in buying it. And he organized it all for us.

'The car was complete, as last raced by Sakai. When we got it, it was just like stepping back into 1974. It was a real time-warp. It was amazing. It was as if Sakai had just stepped out of it. Tony and I raced it at Wings and Wheels [at Whenuapai] two weeks after we got it. We didn't even change the tyres. We kept it for ten years, raced it at historic events all around the country, and took it to Australia once. We kept the [yellow] colours and Minolta signage exactly as it last raced in Japan. We then sold it to Barry Kirk-Burnand.'

Both Fox and Roberts are closely involved with the Bruce McLaren Trust, and Fox went on to restore the 1968 McLaren M8A with which Denny Hulme won the 1968 Can-Am Championship, and which spent much of its life in New Zealand. He has since restored various other Can-Am cars for international customers, and is a world authority on the subject. However, despite selling the McLaren M12, in the back of his mind he had hoped to one day purchase it back, and restore it as the one-off McLaren/Toyota car, as it contested the 1969 World Challenge Cup Fuji 200. But that would require somehow being able to find and purchase a unicorn of the motor racing world; one of the genuine Toyota quad-cam V8 racing engines from the original Toyota Group 7 programme. Incredibly, through Don Nichols in the US, that's exactly what happened. Nichols had established Advanced Vehicle Systems in 1969, and would go on to build and race cars successfully in the Can-Am series and in Formula 1 under the name Shadow. And it was Nichols who held the key to this significant car being reunited with an original Toyota Group 7 engine.

'I always felt this was a project worth doing,' recalls Fox, 'otherwise it was a significant car whose history [with Toyota] was lost. After World War Two and the Korean War, Don Nichols was resident in Japan. He was involved in Intelligence for the US Army. So after he'd served in the military, he stayed in Japan, and he and another guy were responsible for creating Mt Fuji Speedway. He was also responsible for creating the World Challenge Cup Fuji 200. But he left Japan in 1969 and went back to the US, and started Advanced

This was how the McLaren M12 looked when imported to New Zealand by Duncan Fox and Tony Roberts. This is exactly as the car had been displayed in a glass case in a Tokyo restaurant since 1974. It retained its Japanese racing livery.

An incredibly rare original Group 7 Toyota quad-cam V8 engine, which Duncan Fox found in the United States.

Vehicle Systems, where work began on designing and constructing a car to contest the Can-Am series. Through his contacts at Toyota, he got a Group 7 Toyota V8 sent to the US, which was fitted into the first Shadow Can-Am car [the radical tiny Trevor Harris designed Shadow MkI with its 10-inch-diameter wheels and special Firestone-developed tyres]. But he decided the Toyota wouldn't do the job, so he pulled that out and put a big block Chevy in it. So the Toyota engine had sat in Don's warehouse ever since. It's virtually a brand-new engine with about one hour of dyno time.'

Actually acquiring the Toyota V8 took several years. 'I heard a whisper on Facebook about the engine, and I followed it up. And then it took another 12 months before Don would confirm he had the engine. It then took me another 18 months before I could buy it off him. But through that we became good friends. He was a real American hero.'

The engine itself, typically, is a work of art, with virtually everything being completely custom-made. 'At first I assumed it would be a Cosworth [DFV Formula 1 engine] copy. But it's not. It's more like two 4-cylinder motorcycle engines joined together. It's a beautiful piece of kit, and quite exotic. It uses an aluminium block, magnesium oil pan, twin water pumps, one on each side with a scavenge pump on one side which is driven by the water pump, and an oil pump on the other side, also driven by the water pump. The water pumps are belt-driven but all the cams are gear-driven. It's got mechanical fuel-injection, like a Bosch or Kugelfischer, but it was made by NipponDenso in Japan. It's an all-out race engine. It's a pretty exotic piece. I imagine spare parts for it are non-existent. I rummaged through Don's warehouse a couple of years in a row and found a couple of boxes of parts.'

Although an accurate restoration, Duncan chose to make one important improvement; a more appropriately affixed rear wing.

'In my opinion, if that car hadn't DNF'd [Did Not Finish] from overheating, it would have DNF'd from wing failure. The wing was only held in place with two 5/8 inch tube vertical struts, and with a pair of very thin cable lines running from the strut to the roll bar. I'm sure it would have failed.

'The chassis tub is incredibly well preserved, one of the best preserved I've ever seen. Toyota had a month to put the car together and go racing. So it was a bit rough in some areas. The Toyota motor is really too wide for that chassis. It's 6.5 inches wider than a big block Chev. It's huge.

'As I've done more and more research, the more astonished I've become about how technically advanced the Japanese were, and even more so than the US; composites, and stuff like that which the Americans and the British weren't even trying yet. Some of this stuff is really exotic. It's as exotic as any Ferrari or any European race car of the time.'

It seems a pity, then, that Toyota never ventured to the United States with its Group 7 project to contest the Can-Am. It's almost certain, given the ambition of Toyota, and how quickly the Toyota/Yamaha partnership could adapt and progress, that the twin-turbocharged car would have played a significant role in the Can-Am series. This technological marvel should have had its place in American racing history. But it didn't. In fact, Toyota's Group 7 efforts are little-known outside Japan. Despite its potential, the 578A was never raced. Every now and then Toyota will bring out one of its Group 7 sports cars to demonstrate at important racing events or motorsport celebrations, such as the Goodwood Festival of Speed. And understandably, these appearances attract enormous interest.

So we're quite privileged, then, that a small piece of this exciting period in Toyota's history was right here in New Zealand. Duncan finished its epic restoration, before it was shipped off overseas to its new owner.

WELCOME
Hampton Downs

Bob Homewood's Corolla

'I bought a 1969 Corolla to use as a daily driver general runabout around mid-1973. It covered plenty of miles in this role, even more so come January 1974 when I bought my business in Waimauku, northwest of Auckland city. Then it did a round-trip of at least 140 kilometres every day, commuting from Ramarama to Waimauku for a couple of years or so.

'It was dead stock-standard when I bought it from Wrightcars and had seen service as a rep's car in the Wellington area. It covered all that distance plus travelled all over the North Island and also a couple of South Island trips as well. It was dead reliable and never gave us any trouble in all those miles. The only replacements were a clutch plate and brake pads.

'I had been out of motor racing for a couple of seasons mainly due to buying and setting up the business, when the touted Castrol GTX [production racing] series came along for the 1975/76 season. It appealed to me so I set about looking for a suitable car for 0–1300cc class. I did look at buying both an Escort GT and a Datsun 1200 SSS that were competing around the same time. But one day I got to thinking that I want to do something different, so that was how the Corolla became a race car.

'A bit of research confirmed that indeed it looked possible that something could be made out of it, so in due time all the homologated parts were ordered through Toyota New Zealand. I built an engine up into a competition unit, modified the cylinder head myself to the paperwork provided, added a CR gear kit, a limited slip diff, and modified the suspension. One bone of contention looking back was I bought the homologated inlet/exhaust system which was a beautifully built piece of gear from Japan at some considerable cost as I was told that I had to use that exact part. You can imagine my chagrin later when I found that most of the other cars were just using local cheap inlet manifolds and exhaust systems.

'All the cylinder head parts, the CR gear kit and LSD were TRD [Toyota Racing Developments] parts. The camshaft at that stage was a development of my own as was the oil-pump and modified wet-sump. The front suspension in its original form used a combination of a transverse spring and coil springs on the dampers, a system I more or less retained, albeit heavily modified. I found you could end up with a very good progressive spring rate using the system. The front shocks were units I built up and modified myself, the bottom valves being adjustable down through the strut shafts, a system that Koni used some time afterwards although I think my system predated it. Interestingly enough, because of my location out of town, the car never saw a dyno in either of its forms in the time I owned it. All tuning and development was done at the track.

'First time out for it was the opening round of the GTX series at Manfeild in October 1975 where in spite of it only being finished the day before the meeting and being "run in" on the Desert Road on the way down on the Saturday morning, it netted me a third place in Class which was what I was aiming at.

'The next meeting after Manfeild was at Bay Park on Labour Weekend. I was looking forward to it as at Manfeild I had run the small treaded 12-inch-diameter race tyres and had struggled a bit for grip. For Bay Park the 13-inch slick tyres were supposed to be there for us but for some reason they never turned up until after the New Zealand International Grand Prix meeting, so for the next two or three events we just had to persevere using the 12-inch treaded ones.

The Corolla, in its original racing guise, contesting the Castrol GTX Production series.

There was nothing else available. Before the Bay Park New Year meeting I changed the camshaft for one that gave us an increase in power and torque. I also managed to source a set of basically worn-out ex-Formula Ford slick tyres. The combination of the cam and tyres gave us an improvement of nearly a couple of seconds so we enjoyed a much better meeting and were a lot closer to the two front-running cars.

'The Grand Prix meeting was the following weekend, but there was still no sign of the promised new tyres so I decided to persevere with the basically worn-out tyres from the weekend before, which in hindsight was a bad move because in the race I was slowed by a puncture which put paid to another top three Class finish. I had intended doing the Southern part of the series but with no tyres available, and pressure of work, I decided against it. Towards the end of January the tyres turned up at long last. The GTX series was over but I decided to enter a 0–1300 ShellSport race at Pukekohe in February, as I was keen to see how the new tyres worked. In the meantime, there were rumblings that the GTX series would be run to different rules the following season, so I decided as a bit of a toe in the water exercise to see if the car would be competitive with perhaps a couple of modifications in the 0–1300 ShellSport [New Zealand Saloon Car Championship which allowed greater modifications] series the following season.

'The toe in the water exercise proved to be very good. The car in practice was over 2 seconds a lap quicker on the new wheel/tyre combination and returned a quicker lap time than the two front-running [GTX Series] cars did at the Grand Prix meeting. In fact, with a bit of suspension tweaking we later ended up with third quickest time in the ShellSport races. This confirmed my original thoughts that the car would be more than competitive in the class.

CAMPBELL'S HENDERSON
36
SPORT
TOYOTA
TOYOTA

In its ShellSport racing guise, and sporting wide lightweight wheels housed inside MkII Escort bubble flares, the Corolla looks purposeful.

'With this in mind and the impending doom of the GTX series in that form, it was decided to run ShellSport the following season. To that end, further developments were done to the car over the remainder of the season. It was 4-linked with a Watts-link at the rear, wider rims and bigger brakes and a pedal box were added. The car ran an interim trim with some attractive eyebrow wheel-arch flares until they were outlawed. We also trialled a rear diffuser and side skirts but they were also a no-no with the scrutineers.

'Over the winter the car was completely rebuilt, seam-welded and lightened. Mk 2 Escort flares were modified and adapted to fit. I fabricated an alloy air-dam and Steve Carter finished it all off with what I thought was a very attractive paint job.

'Another more powerful engine was built, which used a cam and dry-sump system from Motor Improvements in Australia, and amongst other things, incorporated Cunningham rods and two ring pistons plus bigger carburettors. A 5-speed gearbox was also fitted but we later reverted back to a 4-speed as we found the car was quicker with it.

'The motor was also changed to an upright mounting position in the car. Over the season we ran at lots of meetings and were

Battling the Coppertone sunscreen sponsored Alfa, the famed wide-body 105. Amongst the small-capacity classes, the Corolla and Alfa were two of the more unorthodox racers to combat the popular Ford Escorts and plethora of Datsuns.

generally in the top three. A downside was the amount of panel damage we suffered, most of it as a result of being involved in other people's accidents.

'I had originally once again intended doing the South Island rounds, but business commitments put paid to that again along with a visit I planned to Ruapuna. As the end of the season approached there were once again rumblings that the class was going to be axed, which was a pity as I was just starting to find more time for my racing and another season in the car would have been interesting with me having more time for testing and development.

'The final meeting for me with the Corolla was a dismal wet, complete with thunder and lightning, Club circuit meeting at Pukekohe in May where I trialled the Lucas fuel-injection which I had adapted. The improvement in drivability was impressive. I know we were on the right track. As I said, we just needed one more season.

'At the end of the season I decided to go sports car racing. It was something I had wanted to do for ages and they at least had stability in their rules and to that end the Corolla was traded to Charlie Benseman on his sports car. In turn, he sold it on and it enjoyed a few more seasons with various owners in the South Island.'

Callum Daly’s

Soarer

Toyota created the Soarer as a high-end personal luxury GT, marketed as its 'halo' car. It was created to compete with the Mazda Cosmo and Nissan Leopard, and like those models, wasn't available in New Zealand. With the advent of the import car market, however, the Soarer has become highly prized.

The Soarer spanned three generations, and produced from 1981 to 2005 before being superseded by Lexus. Callum owns beautiful examples of both the second- and third-generation models.

'This is my 1989 Toyota Soarer with a 1JZ engine swap. I owned a JZZ30-shape Soarer before this one which got me into Soarers. To me the Z20 Soarer is the shape you want to own if you like them.

'My white Z20 has been built by two different owners in New Zealand before me who did the engine swap and kept it tidy. Since I have owned it I've replaced turbo seals and just tried keeping it clean.

'I also have a JZZ30 that I've owned for over two years and have done a lot of work on, to make it look exactly how I want it to.'

SOARER OWNERS NZ
JHQ288
I LIKE GOING
SDW4YZ

JHQ288
TOYOTA SOARER

SOARER OWNERS NZ
LITTER!
SDW4YZ

0 1 2 3 4 5 6 7 8
TACH x1000r/min
E F
SPEED
km/h
C H

SDW4YZ
JHQ288

Chris Amon

McCORMACK RACING
PETER STUYVESANT
series
WORLD-WIDE TRAVEL
MOUNT COOK

Chris Amon is often referred to as one of the greatest Formula 1 drivers never to win a Formula 1 World Championship Grand Prix. While it is true, it's selling him short. Grand Prix winner or not, he is rightly considered one of the greatest racing drivers of all time. That he didn't win a Formula 1 Grand Prix, or indeed, a World Championship, is really a result of being in the wrong seat at the wrong time, and pure bad luck.

Amon was born in Bulls in 1943, and went to school in Wanganui. His parents were farmers, and much like the great Scottish racing driver Jim Clark, he himself was born and bred a farmer. When Chris retired from motor racing, he returned to the family farm.

Amon's first car was an Austin A40 Special with which he thrashed around the family farm, and contested local races and hillclimbs. He switched to a 1.5-litre Cooper briefly before purchasing an elderly Maserati 250F, with which he contested the 1962 New Zealand Grand Prix.

Amon raced a Cooper T53 in the 1963 season, but while he regularly qualified well, he rarely reached the finish, the Cooper succumbing to a series of mechanical failures. However, former Grand Prix driver Reg Parnell, who was now running his own Formula 1 team, and brought a team down-under each year for the Tasman races, had been impressed by Amon's showings, not just in the Cooper, but also the outdated Maserati the year prior, later telling the young Kiwi he'd not seen anyone drift a 250F that well since the great five-time World Champion Juan Manuel Fangio.

Parnell convinced Amon to drive for him in the 1963 Formula 1 World Championship. He was just 19 years of age when he qualified his Reg Parnell Racing Lola Mk4A fifteenth for the opening round, the Monaco Grand Prix. However, he was destined not to start the race, as his team-mate, Maurice Trintignant, took over his car after his own developed a misfire. Amon's best 1963 finishes were a pair of seventh places in the French and British Grand Prix. But with points only being awarded for the top six finishers, Amon completed his first F1 season without scoring.

Reg Parnell Racing switched to Lotus chassis for 1964, and although Amon scored his first World Championship points after finishing fifth in the Dutch Grand Prix, his season was beset with mechanical dramas.

Parnell was offered desirable BRM engines for 1965, but on the proviso he took on Richard Attwood as his driver. Reluctantly, he agreed, which meant releasing the young Amon. However, he was soon picked up by fellow-countryman Bruce McLaren, to drive for his fledgling racing team. Amon scored the first race win for McLaren's new M1A Group 7 sports car in early 1965, but otherwise, this was to be a quiet season, although he did a great deal of testing for tyres which would stand him in good stead later.

In 1966, Amon drove alongside Bruce McLaren in the newly established Can-Am sports car series in the US, where he finished sixth in the points standings, and regularly qualified in the top three.

But while it had its frustrations, 1966 also produced some important results. Bruce McLaren had been working closely with the Ford Motor Company, developing the GT40 endurance sports car. McLaren and Amon drove GT40s at Le Mans in 1965, teamed with other drivers, but both failed to finish. Ford stepped up its game for 1966, fitting the car with a big block 427 cu in engine, now named the MkII. McLaren and Amon were paired together in a MkII for the Daytona 24 Hours race, where they qualified seventh and finished fifth.

Chris Amon is best remembered for being a Ferrari factory team driver in the late 1960s. As well as racing Ferraris in Formula 1 and World Sportscar racing events, he also raced this beautiful Ferrari 246T in the 1968 Tasman Series. These photographs are from the Teretonga race.

Ford then put McLaren and Amon together in a black and silver Shelby American prepared MkII for the Le Mans 24 Hours race. In a Ford-contrived 1-2-3 formation finish, the Kiwi drivers took victory, from the team cars of fellow New Zealander Denny Hulme and Ken Miles, and Ronnie Bucknum/Dick Hutcherson.

Despite his series of frustrations and false-starts in Formula 1, Amon had done enough to impress Enzo Ferrari. He signed on to drive Formula 1 and Sports Cars for Maranello in 1967. In F1, Amon's team-mates were Lorenzo Bandini, Mike Parkes, and Ludovico Scarfiotti. At Monaco, Amon performed well to finish third, behind Denny Hulme and Graham Hill. However, celebrations were muted in the Ferrari camp, as Amon's team-mate Bandini was killed when he crashed and rolled his car, and was trapped beneath it as it exploded into flames.

During the rest of the season, Amon achieved four more points – scoring results but was too often let down by his machinery.

Amon also performed World Sportscar duties in 1967, with sports car racing being of great importance to Enzo Ferrari. Highlights included victory alongside team-mate Bandini in the Daytona 24 Hours, ahead of the sister car of Parkes/Scarfiotti. Amon and Bandini also won Round 3 of the championship, at Monza.

Amon worked closely with Ferrari designer Mauro Forghieri, who was highly impressed by the young Kiwi. The 1968 Ferrari 312 was fitted with aerofoils, which were tested successfully when the team sent a special 2.5-litre car to New Zealand for Amon to drive in the Tasman Series.

With two race wins and a number of podiums, Amon finished the championship in second, to Jim Clark, driving the impressive new Lotus 49 which had been debuted partway through the 1967 Formula 1 World Championship.

Amon and Ferrari appeared one of the few combinations who looked like they could challenge Lotus in the 1968 Formula 1 World Championship. However, this would prove to be a year of frustration. In the 12-round championship, he managed to greet the chequered flag just four times. He retired seven times, while Ferrari skipped the Monaco Grand Prix. His best finish of the season was second in the British Grand Prix. Here he battled with eventual winner Jo Siffert, and only narrowly missed out on the win.

1969 started well for Amon, when he returned down-under as part of a two-car Ferrari team for the Tasman Series, alongside Derek Bell. He qualified on pole position for the New Zealand Grand Prix, and went on to win the race for the second year running. Then he won again at Levin. He was third in the Lady Wigram Trophy, and third at Teretonga. He won the Australian Grand Prix, held at Lakeside, but crashed on the opening lap at Warwick Farm. The 1969 Tasman Series only had seven rounds, the last of which was held at Sandown, where Amon took his fourth win, and was crowned 1969 Tasman Series Champion.

In Formula 1, with its 3.0-litre-engine rule, Ferrari used its V12, and while powerful, it was chronically unreliable. In the first six races, Amon finished third in the Dutch Grand Prix, and failed to finish the other five.

Amon also drove the team's sports cars, and banded together a small team to run a Ferrari Can-Am campaign in the US, which resulted in a monstrous 7-litre V12 engine, the largest motor built by Ferrari to date. However, it couldn't compete with the dominant McLarens, which won all 11 races in 1969.

Ferrari had spent much of 1969 developing a new flat-12 engine for Formula 1, and while it proved fast in testing, Amon had lost confidence in the company's ability to

Top: *Amon raced this March 701 during the 1971 Tasman Series.*
Above: *During the Teretonga round of the 1969 Tasman Series, Amon and team-mate Derek Bell chase the Lotus 49B of Graham Hill and eventual race winner Piers Courage in the Brabham BT24.*

provide a reliable 12-cylinder engine. Since its arrival in 1967, the brilliant Ford Cosworth DFV V8 had proven the best power unit on the grid. Furthermore, anyone could purchase one. By 1969, most Formula 1 cars had a Cosworth DFV nestled in the back. Amon wanted to join them.

In 1969, a new operation was established, called March Engineering. For 1970, they were to produce the chassis for reigning World Champions Tyrrell, whose lead driver was Jackie Stewart. March also ran its own factory team in 1970, with a pair of Robin Herd designed cars designated 701, and took on drivers Amon and Jo Siffert. STP was to support Amon, while Porsche paid for Siffert's seat. In addition, Amon would race a massive STP-sponsored March 707 in the Can-Am series.

A new team, and optimism for a successful season. But 1970 would be much like Amon's three years with Ferrari. Amon finally reached the finish in Round 4 at Spa Francorchamps, in the Belgian Grand Prix. He took a close second to Pedro Rodriguez, and also broke the lap record. He retired from the Dutch Grand Prix, but finished second again in France.

Tensions rose between Amon and Mosley after the British Grand Prix, as the Kiwi claimed to have monies owing from March, and was reportedly supplied with weaker DFV engines. To add to his woes, after three years of frustration at Ferrari, the Scuderia finally came good in 1970, with its new flat-12 winning four of the last five races.

For 1971, Amon switched to the French Matra team, who'd entered Formula 1 in 1968. But this would be another tough year, with a single third place the only highlight. 1972 was much the same. He scored his last F1 pole position in the French Grand Prix, where he finished third, his best result of the season. After Matra withdrew from Formula 1 at the end of the year, Amon switched to the new Tecno outfit, but 1972 was a disaster, and he left the team with four races remaining.

In 1974, Amon ran his own team, but the Gordon Fowell designed Amon AF101, while a handsome-looking car, lacked development. He returned to race the Tasman Series in 1975, which by now was being contested for Formula 5000 cars. Driving a Talon TR1, he qualified on the front row three times, and won the Teretonga round. Back in Europe, he raced a handful of Formula 1 events for the Ensign team, but failed to score any points. He again returned with Ensign in 1976, and showed he was still one of the fastest drivers on the grid, qualifying third in the Swedish Grand Prix, and looked to be on target for a podium, until the suspension failed. When Niki Lauda suffered his fiery crash on lap 2 in the German Grand Prix, Amon refused to restart, having seen too many of his colleagues killed or injured during his career. He was fired on the spot.

In 1977, Amon quit the sport for good. He was 34 years of age, and had been living and racing in Europe since 1963. He was happy now to return to New Zealand to run the family farm.

In the early 1980s, Amon appeared on a New Zealand television show called *The Motor Show*, hosted by presenter and car enthusiast Dougal Stevenson. In it, he test drove various new cars around his local race track, Manfeild, and gave his opinion on each. Toyota felt that when Amon tested their cars, he was particularly critical. This created grumblings within the company, and feelings of resentment, that Amon was being particularly unfair on their product. Some suggested ignoring the criticism. But Bob Field, who was in charge of Toyota New Zealand, instead took a pro-

STP
AA
STP

active approach, and invited Amon into the company office for a discussion. On spending time together, Field challenged Amon to help develop Toyotas into better cars. Amon accepted.

The relationship was a happy and particularly successful one. Toyota, who were still a relatively young company in New Zealand, now had a world-famous racing driver developing its cars for the local conditions. So successful were the results that Toyota Japan invited Amon to spend time developing their vehicles. However, he didn't want to spend long periods of time in Japan. So they agreed to bring their development programme to New Zealand.

Toyota also established Toyota Team New Zealand, running Celicas and Corollas in production racing. Amon was involved with this programme. He emerged from retirement to drive a Toyota Camry in the 2003 Dunlop Targa, with famous Formula 1 television commentator Murray Walker as his co-driver. He also contested the 2004 EnergyWise Rally, driving a Prius. He won. When the Toyota Racing Series was established (see pages 174–185), in an effort to give young Kiwi drivers local experience in a high-intensity wings and slicks single-seater formula, Chris Amon was the class patron. Among the highly desirable trophies on offer for the various races, including the New Zealand Grand Prix, Lady Wigram Trophy, New Zealand Motor Cup, Denny Hulme Memorial Trophy, and Bruce McLaren Trophy, perhaps the most highly coveted is the Chris Amon Trophy, as this is awarded to the driver who has accumulated the most points throughout the series. That person is the Toyota Racing Series Champion.

Chris Amon passed away in August 2016, aged 73. He remains, to this day, one of the most celebrated racing drivers of all time. Internationally, he will always be associated with Ferrari. But, in New Zealand, he'll always be associated with Toyota.

Chris Henderson's GT86

New Zealand continues to enjoy a long history of wildly creative racing cars. From the Specials that emerged during the 1950s to take on the world's best Formula 1 cars, to the Allcomer saloons in the 1960s, the South Island based OSCA (Open Saloon Car Association) modified saloons of the 1970s, and the Sports Sedans of the 1980s, conjuring up incredible innovations just seems to be part of the Kiwi DNA.

Endurance racing is a form of motorsport competition that has enjoyed rapid growth in recent years. Much like the 1950s golden era that pitted some of the world's greatest F1 cars against the most ingenious concoctions local racers could muster, so this fast-growing formula pits the best European and American factory GT racing machinery against some wonderfully creative Kiwi Specials. And Chris Henderson's incredible Toyota GT86 is surely one of the most awe-inspiring of the local breed.

Chris has a long history of racing Toyotas, which have ultimately led to his GT86, with which he contests both the South Island Endurance Championship, and OSCA.

'I started racing cars at the age of 18 with an AE86 Trueno that I drove to and from work and raced on the weekends. It had a standard 4A-GE in it. I then put a half cage in it and circuit raced it for many years. Wayne Murdoch and Bill McFarlane helped me a lot with this car and I raced the endurance series with Bill. I also won the Sport Saloons Group 2 series with this car in 1992.

'I then went rallying in a Mazda 4WD rally car, and it nearly broke me. I also had a Mazda RX7 Classic rally car, now owned by Mark Leonard.

'Then I owned a Group A AE86 circuit car, built originally by Brian Bates in Palmerston North. This car finished eleventh overall and second in class at Bathurst in 1988. By the time I got it, it had been well used. The Quaife gearbox was ready for a rebuild and the motor was well below average. I raced it in Sport Saloons but didn't have the horsepower to be competitive. I turbocharged it for the next year and won the Sport Saloons overall championship.

'Then I built an E36 BMW with a Nissan RB26 motor. It was a fast endurance car but then the global financial crisis hit and I sold it. I took a couple of years off and then built a Toyota AE86 with a Chevrolet LS3 in it. We won the 1 Hour South Island Endurance Racing Drivers Championship (SIERDC) Championship twice in this car until an incident at Teretonga saw it destroyed. It was then decided to put the running gear into a later model vehicle.'

This is where Chris's current GT86 enters the picture. It wasn't an 86 series one-make race car, but rather a damaged road car. 'The body shell we found in Blenheim. It was due to be wrecked. It had a slight accident in the rear that didn't affect what we were going to do with it, so we purchased the rolling shell only.

'We fitted it with Nissan Silvia S15 suspension in the front, and removed all the rear structure and panelling except for the chassis. We used the Ford 9-inch race diff out of the old [AE86] car and fabricated the mounts into the GT86; as they are all independent rear-end we had to clearance some metal.

'We removed the firewall completely and sat the engine as far back as OSCA rules allowed. We run the factory suspension pick-up points in the front. The front brakes are AP 6-piston endurance front calipers. These take the larger brake pad so don't suffer from heat so much. The rotors measure 385mm front, and 330mm rear.

'The engine is a 6.2-litre LS3 with Kelford cam and Jenvey throttle bodies. It makes around 600 horsepower. It's backed by a TT Industries 6-speed sequential gearbox. The Ford 9-inch race diff uses a Detroit locker.

The GT86 as purchased by Chris as a damaged road car.

'It rolls on 18 x 11-inch wheels with Kumho slick tyres. It has no ABS, no paddle shift, no auto throttle blip, and no traction control. It is a real driver's car. It has a 52/48 weight split and weighs 1200kg.

'Phil Winter designed the cage and rear suspension mounting. I did all the internal panelling and finishing with a mate of mine, Graeme Wilson. The assembly was all carried out by my team of guys at work [Midway Motors in Dunedin]. The engine work and tuning is by Mike Lynn of the Turbo Shop in Dunedin.

'We built the car to basically replace my AE86. It competes in the South Island Endurance Series 1 Hour, OSCA (we won the championship first year out), and it's a good club car. We kept it within the OSCA rules so it is not an outlaw car and always has somewhere to race.

'My plans going forward are to have some fun with the car. I'm a little bit over the intense stuff, championships and race series, for now. I have won the SIERDC series twice, and OSCA. When you build and run your own car it is hugely stressful. I'm enjoying a break at the moment so will see how long that lasts.

'As for future upgrades, none are planned at the moment. I like that it is raw and needs to be driven. I believe all the driver aids nowadays are taking away from the driver.'

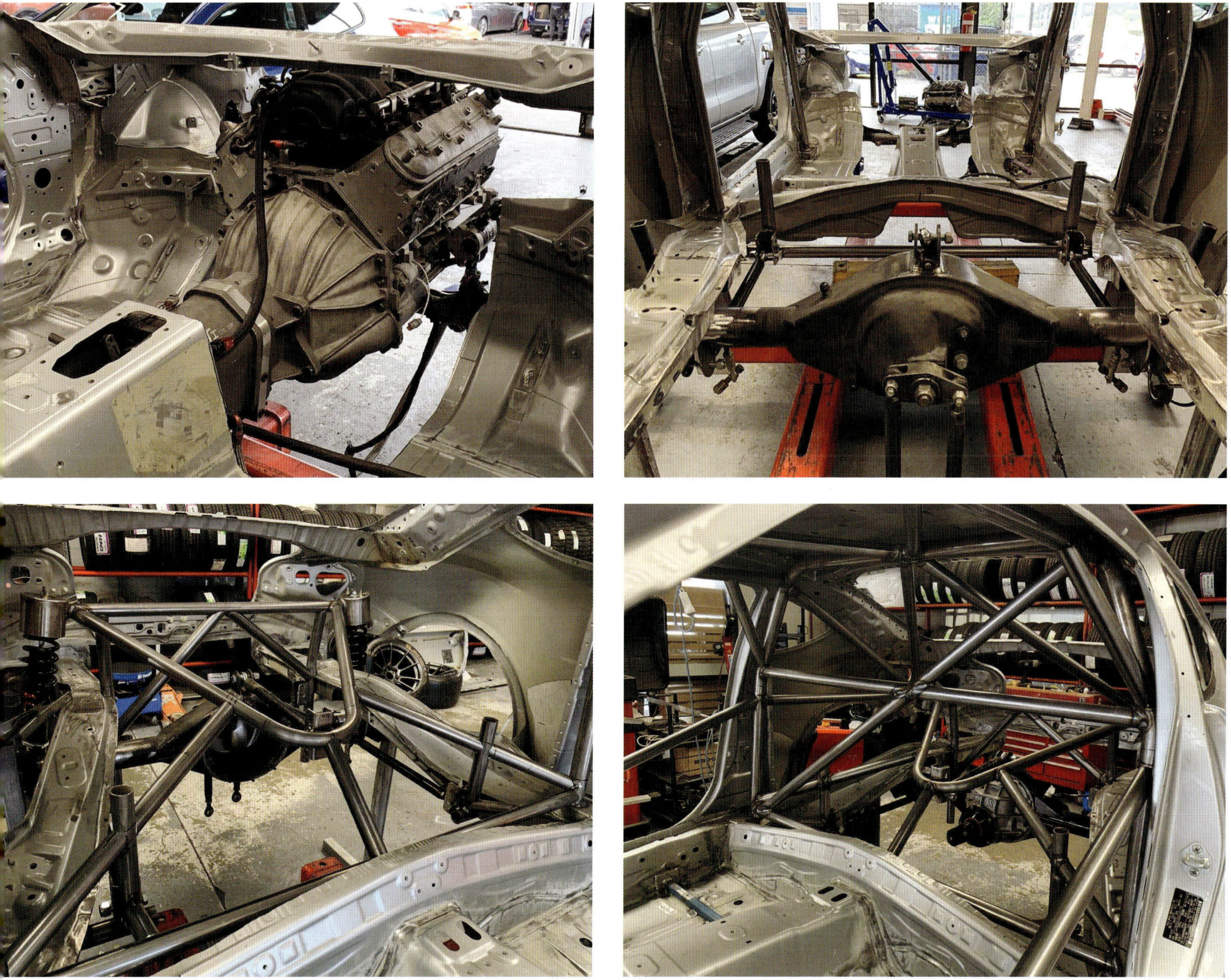

During construction, and the level of fabrication work required was extensive. The big Chevy V8 fills a significant amount of space.

Chris's Chevrolet LS3-powered Toyota AE86, which, following a heavy shunt, donated much of its running gear to the GT86.

52
midway motors

52
CARTER'S tyre service
Allied 24 7 Fuel
PFC BRAKES
MoTeC
PFC BRAKES
MoTeC
Turbo Shop

In amongst the high-dollar Porsches that are commonplace in endurance racing, Chris's wild GT86 holds its own, both in terms of aesthetics and performance.

Daniel Gondrian's Celica

The Celica was Toyota's answer for the 2+2 sports car market, and ultimately spanned nine generations before the last example rolled off the assembly line in 2006.

Throughout much of its life, the Celica also was Toyota's chosen model for motorsport competition, beginning with the first-generation model which competed both in touring car racing on the tarmac, and in rallying. Indeed, rallying had become a major component of Toyota marketing by the 1990s, when the sixth-generation model was launched. By then, FIA Group A was the formula for the World Rally Championship (WRC), and Toyota were fully committed.

Carlos Sainz won the WRC for Toyota in 1990, and again in 1992, while Juha Kankkunen won in 1993, and Didier Auriol in 1994. All four championships were won using the Celica. To achieve the success it did, Toyota designed and built the GT-Four. For the model to contest the WRC, Toyota was required to homologate it, which involved producing a limited number of road cars. For homologation purposes, a minimum of 2500 cars were required to be built in a 12-month period, although this increased to 5000 cars from 1993. In addition, manufacturers could develop and sell an 'evolution' model, based on the mass-produced examples, but sporting specific competition components to be carried across to the race cars. At least 500 evolution models were required for the components to be homologated.

Today, clean road-going GT-Fours are highly desirable, and Daniel's example is one of the nicest anywhere. 'This is my 1996 Toyota Celica GT-Four ST205 aka "Flaps". I bought her a few years ago for my 21st birthday, because who doesn't buy themselves cars for their birthdays!

'I have had a few GT-Fours so far but this one by far is the best. She has had some engine modifications done so she loves to boogie a lot. I've done so much with this car, I've driven her to Auckland and back twice so far. I've driven all over the show in the South Island. I've raced her around the street sprints in Ashburton and she just keeps going.

'This is a car with such a big personality and such character, it's just one amazing feeling every time I pick up the keys and start her up. Toyotas in general are amazing but this girl right here is irreplaceable. I can't get enough of the GT-Four.

'This is a very underrated sports car that not many people know about. I love when people come up and ask if she's a real GT-Four, because once I say "yes", people's faces drop and they're amazed because there's so few of them around. It's amazing.'

Edward Wong's MR2

At the heart of Toyota is a passion for driving enjoyment, and as such, the company has invariably had a focused sports car in its arsenal. But few have been as dramatic as the MR2.

The MR2 were sold from 1984 to 2007, during which three generations of the model were produced. What made it so unique was that this was a genuine two-seat mid-engined sports car, with the motor mounted behind the occupants, but ahead of the rear axle line, much like a Formula 1 or World Endurance Championship sports racing car. It was, in effect, a pint-sized affordable supercar.

The name MR2 derives from its layout; Mid-engine, Rear-wheel-drive, 2-seats, or Mid-ship Runabout 2 seats, depending on your preference.

In 1986, Toyota introduced a supercharger engine option, which bumped horsepower from around 115 to 145. And Edward is the lucky owner of a first-generation, first-year supercharged model, a 1986 AW11 MR2 Super Charger, 1600cc twin cam 16v 4A-GZE.

'I suppose you could say my interest in first generation (AW11) MR2s began at an early age. The first time I ever saw one, growing up in the 1980s, I mistook it for some kind of Ferrari, because that's what typically passes for cool when you're five years old.

'My parents still have some childhood drawings stashed away of "Ed's dream garage" where one of the cars has "MR2" proudly scrawled in the space between the tail lights. It even looks like one, but maybe that's because the sharp, angular styling made them incredibly easy for any average eight year old to draw. By then I knew it was some kind of Toyota, and I knew it was pretty unique, because it only had two seats, and the engine was in the back. Sort of like a Ferrari.

'So it was inevitable that many years later multiple MR2s would eventually find their way into my possession. This is the sixth one I've owned, privately imported from Japan after some years searching for a tidy rust-free example to hold on to.

'Of the many different permutations available, I've owned them all; normally aspirated, supercharged, T-top, hardtop, manual, automatic... and to me, the pick of the bunch has to be the supercharged manual hardtop. The amount of punch you get from the supercharged engine is already impressive for such a light chassis, so I haven't found too many excuses to increase the power level beyond rational. Under the engine cover the only fettling that's occurred is in the form of a free-flowing exhaust, extractors and a minor supercharger pulley upgrade. End result is an extra bit of shove in the back when darting around the hills in third gear, with the whine of the supercharger just behind and to the right of your head adding itself to the mix.

'The licence plate is pretty self-explanatory. Everything about the MR2 screams '80s icon, and as a fan of the era I totally dig it; from the aforementioned angular styling to the cosy fighter pilot-style cockpit that could have been lifted straight from the set of *Star Wars*, and let's not forget those trademark pop-up headlights! Because of this I tried to keep the appearance of the car as faithful to the 1980s as possible. The wheels are Bridgestone race alloys [not steels!] from the TRD parts catalogue circa 1988. The car rides slightly lower on TRD shocks and springs, and the funky roof racks are a super-rare factory option that I was lucky enough to find ex-Japan. As for the eye-watering Recaro seats, let's just say they're a nod to the strange fashion choices of the time...

'As a personal collector of 1980s Japanese sports cars [my modern day "dream garage"], the MR2 holds a certain pride of place amongst the rest of the fleet, being the only mid-engined and supercharged car of the lot. You only need to drive one, preferably on a twisty scenic New Zealand back road, to see why people raved about them back then and still do today. In terms of grip, chuckability and responsiveness, it's pretty much the best all-rounder of its time [admittedly my RX7 does come close]. Back then everyone thought Toyota had gone mad for defying tradition and producing a small mid-engined two-seater, but it was a gambit that's definitely stood the test of time.

'Growing older I've found my mindset shifting away from racing cars to restoring them instead. This has coincided with 1980s cars becoming scarcer but more sought after amongst people my age. So I hope that in the years to come there will still be a few AW11 MR2s left for a new generation of fans to appreciate.'

TOYOTA
80S KID

80S KID

TOYOTA
80S KID
RANGIPO
POWER STATION

れたシェイプドボディー。自らの走る姿を、見たくなるほどに美しい。
SUPER CHARGER
MR2 'G-Limited Super Charger,

Eli Gifkins' Ministock

Ministock racing was introduced into speedway as an entry-level, budget formula, originally aimed at young racers. Initially, it was designed as a non-contact class, but was ultimately opened up to allow a separate Ministock division for adults, albeit using the same cars. Light contact is accepted in the adult races.

Ministocks are powered by small-capacity Toyota and Datsun motors, and are designed to be cheap to run and maintain. This particular car was raced by Eli Gifkins out of Palmerston North Speedway, as a first step towards moving up to the full-contact Stockcar class. He began racing the car at age 11. The donor car was a Toyota 1200cc 3K.

The chassis was built by Rees Race Cars and raced by Asher Rees as 11P in the 2006/2007 season. It changed hands several times and was raced by Zane Dykstra, among others.

The car was purchased by Jack McLauchlan and was raced in Nelson by rookie drivers during the 2016/2017 season, as 9N with Jack mentoring and teaching the art of speedway. It was then purchased by the Gifkins family in early 2017.

Eli began racing it during the 2017/2018 season at Palmerston North, Wellington, Wanganui, including the Oceanview 5000 and at Ministocks in Paradise, in Rotorua. He has since gone on to bigger and faster machinery, making the switch to Stockcars in 2021.

Gerard Richards' *Sprinter*

Gerard Richards is a celebrated Kiwi motoring scribe, with a passion for classic cars and motorsport history. He is also a lifelong Toyota enthusiast, having owned several examples. But his most prized was a 1983 Sprinter, a Japanese import sporting the brilliant and long-running 2T-GEU twin-cam engine, being one of the last models to do so before the introduction of the 4A-GE.

'This maroon 1983 imported Toyota Sprinter Twin-Cam, came into my hands sort of by default. I'd gone out to the Manukau Car Fair [in South Auckland] sometime in 1995, looking for an R30 Nissan Skyline Coupe and came back with the Toyota. I'd passed up on a rare 1985 RS2000 DOHC R30 Skyline Coupe, baulking at what I thought at the time was an excessive price, but which in hindsight was a bargain.

'The '83 Sprinter caught my eye instead and it came home with me. I didn't really appreciate its pedigree for some time, as I was smarting at my poor decision on the Nissan.

'What I didn't realize at the time was that fortune had bestowed on me the final and ultimate performance version of the E70 Corolla/Sprinter [produced from 1979 to 1983]. My car, the last of that line, had the 1600cc 2T-GEU twin-cam 8-valve motor with the superb Yamaha aluminium/alloy head. Nowadays this version of the E70 model is the most sought after...

'It didn't take me long back in the day to have my attention fully engaged to what an amazing powerhouse was under the hood. The bodywork wasn't that flash by then, but that engine was a horsepower factory! It thrived on revs and just wanted to keep winding out, with a super-smooth 5-speed gearbox to match.

'My lasting memory is that it was such a fun car to drive. The

twin-cam motor was a beautiful engine with a lovely exhaust note. Despite being the last of the rear-drive versions, it put the power down well and was a rocketship away from the lights on a dry road....

'A major contributor to the E70 Sprinter's performance, were the improvements made to the suspension with this model. The crude, leaf-sprung rear axle that had been the mainstay of all Corolla versions since 1966, had just been disposed of for this model. It was replaced with a four-link and lateral rod setup. The new suspension was groundbreaking and transformed the Corolla/Sprinter into a sharp handling tearaway.

'My tenancy with the little asphalt screamer wasn't that long. The growing issues with the bodywork, or more accurately, increasing volume of substitute metal filler, was becoming worrying.

'I briefly had thoughts, that saving that magic engine/ transmission, to replant in something else could have been an option. But it sold fairly quickly. Someone must have known more than me about its potential future value...

'In hindsight (a wonderful thing), given the value that some of these rare survivor cars like the E70 Sprinter are realizing today, it would have been astute to find a shed for it. But we were ignorant of their quality and the historic part they played in the Japanese heritage performance story back then. Like my 1983 Corolla E70 Sprinter, these banzai equipped pavement scorchers were totally different animals to the local offerings.

'The sad truth is most of these early neck-snapping ballistic Japanese performance cars of the '80s and '90s imported to our shores succumbed to abuse and neglect, or saw out their final days on the dirt speedway ovals. I guess we just didn't realize their classic value at the time.'

TOYOTA SPRINTER
NY1539

NY1539

Glenn Holmes' *Hilux*

For a short time back in 2015, this 1993 Toyota Hilux was a minor celebrity. It was owned by my father, Glenn Holmes.

Glenn purchased the Hilux brand new, and his haulage job around the Manawatu ensured it clocked up a significant amount of kilometres. It was travelling an average of 300km per day, six days a week. At around 1800km per week, that equated to over 93,000 kilometres per year.

In 2006, the Hilux clicked over 1 million km. At that time, it was placed into semi-retirement, and Glenn purchased another vehicle to do the grunt work, while the Hilux took on a part-time role. By 2015, it had clocked over 1.2 million kilometres.

While Toyota vehicles are renowned for their durability, Glenn's rigorous maintenance schedule played a significant role in his Hilux's longevity. A mechanic by trade, Glenn religiously changed its oil on the first Monday of every month, from 1993 until 2015. Impressively, it still has its original engine, which has never been removed, the cylinder head has never been off, and the diff has never been pulled.

In 2015, Glenn decided it was time to part ways with the Hilux. He sold it to Turners Car Auctions, who placed a $1 reserve on it. In the end, it was purchased by a Christchurch bidder. It was the highest mileage car Turners in Palmerston North had ever sold.

The high mileage of the Hilux drew the attention of the media, and Glenn, along with his old work-horse, featured in a TV3 news story.

Isaac Western's MR2

Toyota's MR2 was first launched in 1984 and, as such, is nearing 40 years old. The first-generation model fully encapsulated 1980s automotive design, in that it featured straight lines and sharp edges, only letting up with the wheel arches.

These fun little wedges are pure driving machines, and these days have a very strong following among enthusiasts, and are also commanding strong resale value. Finding a nice, unmolested example is not easy, and those who are fortunate enough to own one tend to cosset it, with good reason.

Isaac Western has had a lifelong passion for the model, and owns this beautiful 1988 example. 'I've been involved in MR2s for almost half my life, which is thanks to/the fault of my dad. He's been through a number of AW11s and a couple of SW20s — introducing me to the MR2 world.

'For a while before I was old enough to drive, my dad was using this very car — a midnight blue, manual, 1988 hardtop Supercharged AW11. It was largely stock, and in good condition. Eventually, the 4A-GZE's front oil seal went, so the engine was removed, fixed up, and put into another one of his AWs. The engineless chassis ended up sitting as a lawn ornament for a number of years.

'After learning to drive and getting my full licence in an SW20 MR2, I wanted to find an AW. My dad and I looked around for a good while but most of the ones we found were terminally rusted from living out in the rain. Long story short, stuff happened, and I ended up buying Dad's rolling chassis off of him — with his promise of helping me get it back on the road. Nothing happened for a good year and a bit, but we slowly got the things we needed together.

'The summer of 2017 is where things actually started progressing engine wise. We ripped two late 80s big port 4A-GZEs apart, and put the best condition parts from both aside for one engine. We had the block and head decked, while we cleaned up the pistons, rods, crank, valves, cams, intake, blah blah blah — everything basically.

'This was the first engine I'd ever pulled apart fully, and I can safely say I learnt an absolute tonne disassembling and reassembling everything with my dad's help. A textbook father-and-son project.

'We cleaned, restored, and handled basically every part on this engine — apart from the SC12 supercharger; we knew the history of it, and it was in good condition.

'To keep costs down, we kept the engine completely stock. This is fine by me, because it's almost exactly the same as when it came from the factory in 1988. It also gives me a solid platform to upgrade and learn further from.

'We did the rings, bearings, and 90% of the seals and gaskets. Along with a new waterpump, dizzy cap, rebuilt alternator, and fresh clutch, the mechanicals should last me years.

'I've done about 2500km since we got it running, so it's all run in and on proper oil now. I try to thrash the great backroads we have in Auckland as much as I can... It might be a 1.6 litre, but when the supercharger is on, it knows how to drink gas. Oh well, whatever the price of fuel is, I'll hopefully keep this car for a very long time. It has a large sentimental value to me, and AWs are appreciating in cash value to others too. Maybe I'll have to get rid of it in the year 2042, if I accidentally total it as a 500hp time attack car or something. We'll see.

'I love the way AWs handle; the fact that they're mid-engined, and have no driver luxuries like ABS or power steering really puts it down to the driver to get the most out of them. Most AW owners have

said this at some point: they handle just like a big go kart.

'It's only about 6.5 seconds to 100 — which isn't fast — but it'll keep up with most cars on a backroad easily.

'I also like the way they look [for the most part]. I reckon the front end could do with some adjustment, but inside and out it really makes it obvious that it's a sports car from the 1980s. Not too out there for the '80s, but definitely '80s.

'Another thing I like is the fact that they're pretty unique and just strange. "A mid-engined supercharged Toyota? Is that a real thing?"

'There is still a lot of stuff to do to this car. For a start, I'm constantly on rust patrol; there are a few small spots of rust that need dealing to, but they aren't structural, and aren't too big. It's in pretty good condition rust wise, but you can never let your guard down... There are small things like changing the gear knob that I don't like the look of, the broken interior light, and finding a cool cassette head unit that looks '80s enough (it has nothing currently. I'm going to use an Aux cassette tape).

'Future plans are to sand and repaint the bonnet that has a giant white stripe down the middle for some reason, the faded engine lid, and super-scuffed chin spoiler.

'Those are small and relatively cheap things. Expensive things in the near future involve some nice 15-inch wheels, and an overboost pulley to up the boost to 14 psi. That should help it keep up with most '90s turbo cars...

'Anyway, yeah nah that's my car. I wrote too much, but yeah. It has a bit of a story behind it I guess.

'Also, while you're here, I took all of these photos! I can do the same for your car if you want hahaha. Check out @Rotationals on Instagram if you want to see more. Self-promotion in Steve's book... Cheeky or what?'

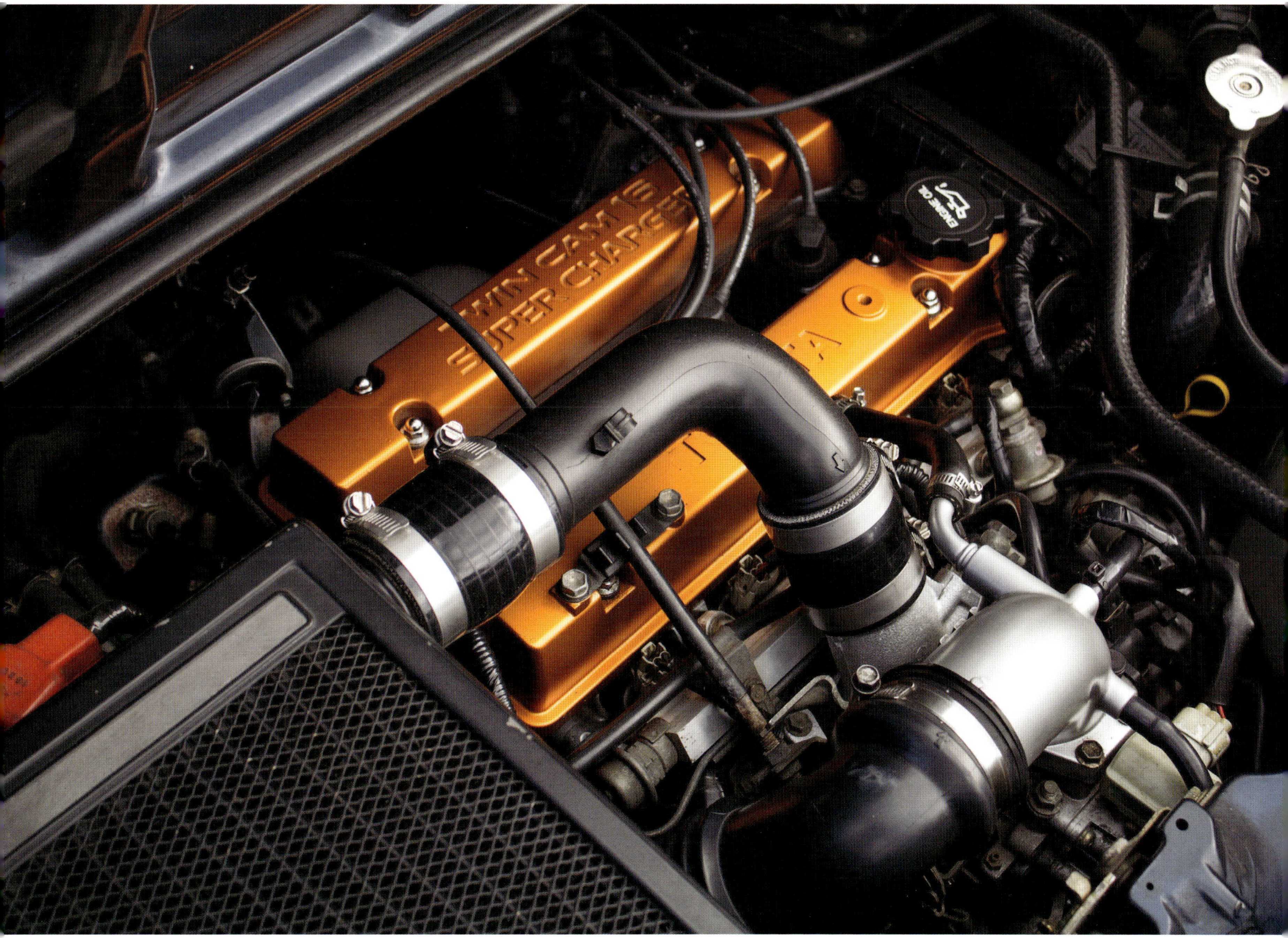
TWIN CAM 16

Ivan Swain's *Corolla*

'This is not an unusual car, but I hope it's of interest. This was my late father's car. His name was Ivan Swain and he worked as the postie at THC, Milford Sound.

'This photo was taken in the mid-1970s on the eastern side of the Homer Tunnel, at the start of winter.

'My father won the money to buy this car from betting on the horses. I think he won around $2000, which, in the early 1970s, was a lot of money.

'He also traded his 1962 Morris Mini 850 as part of the deal. I'm not sure what the total cost of the Toyota was.'

Keith Swain

HY 7938

James Horner's Corolla

Toyota's third-generation Corolla, first introduced in 1974, is one of the great automotive designs, particularly in the now very rare two-door coupe body style.

The model adopted the popular (for the time) 'Coke bottle' styling, with its kicked-up rear quarters, and swooping lines.

Back in the 1980s, third-gen Corolla coupes were hugely popular among young Kiwi street racers and cruisers, whose first modifications were to lower the suspension, fit a loud exhaust, and bolt on a set of Hotwire mag wheels wrapped in embossed white-letter tyres. And it was in this guise, many met their demise. Back then, they were plentiful and relatively cheap.

So it's always exciting to see an original survivor, sporting period modifications, like this beautiful example owned by James Horner. The Corolla didn't look this good when James first purchased it. Its current condition is all down to him, and his enthusiasm for the model.

'I have a few old Toyotas, this is one of them. It's a 1977 Toyota Corolla SR Coupe. It's travelled 90,000 kilometres from new.

'I've owned it for about six years. I had panel and paint done when I got it and rebuilt the motor. I lowered it and got the wheels for it.

'Since then I've driven it to Auckland and back, and to Christchurch twice for car shows. I just love taking it on road trips. I also have a Toyota Crown that's in the build.

'I've also collected old Toyota books, sales brochures, and manuals.'

SE
TOYOTA
TOYOTA

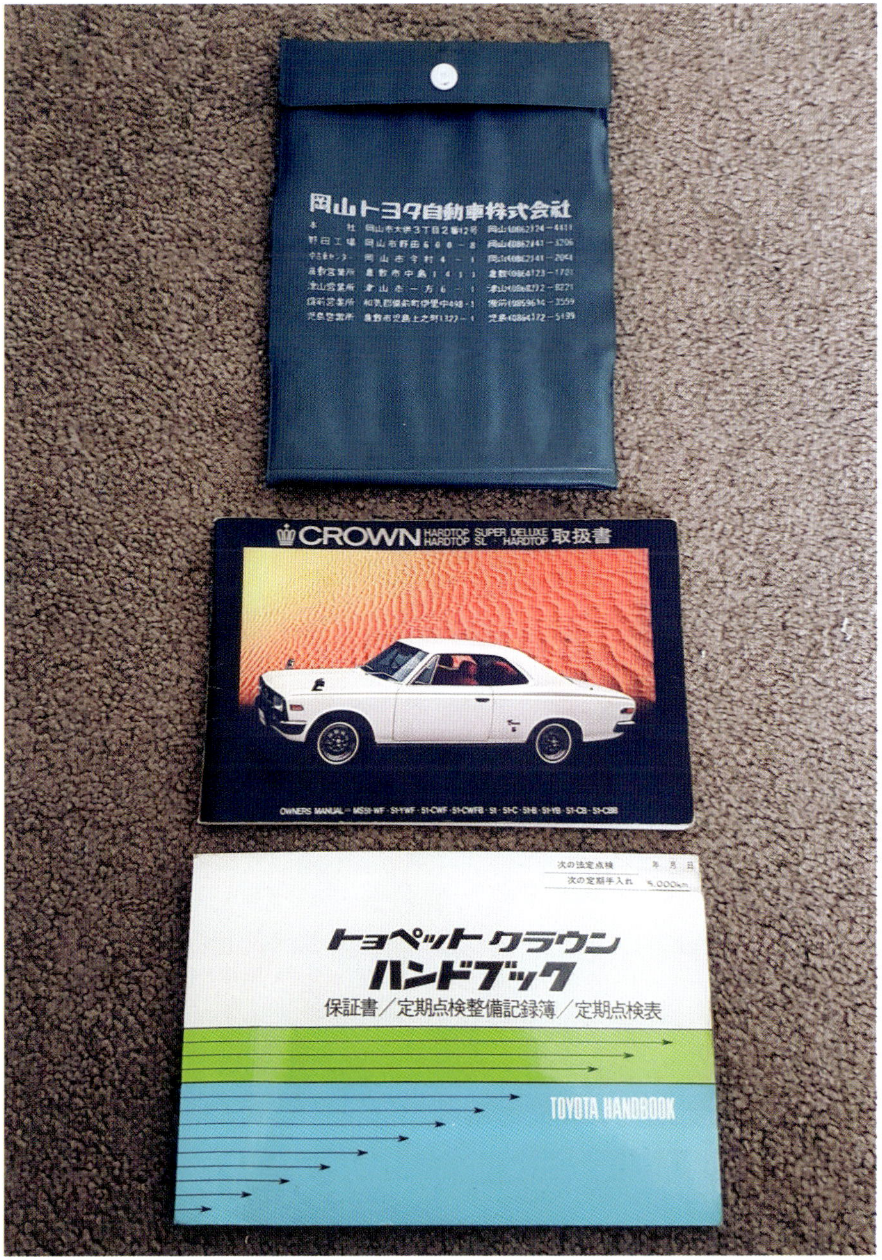
岡山トヨタ自動車株式会社
CROWN HARDTOP SUPER DELUXE HARDTOP SL · HARDTOP 取扱書
トヨペット クラウン
ハンドブック
保証書／定期点検整備記録簿／定期点検表
TOYOTA HANDBOOK

Jayne Hirst's *Celica*

The fifth-generation Celica was launched in 1989. While the styling of the fourth-generation model was still evident, Toyota designers fully embraced the new 'jelly mould' body shape that would take the 1990s by storm. Indeed, barely a straight line can be found anywhere.

Three body styles were offered: notchback, liftback, and convertible (Jayne Hirst is the proud owner of one of these). And hers is an absolute stunner.

'A friend saw this car on TradeMe advertised with a blown head gasket,' explains Kevin Hirst, Jayne's husband. 'He was interested in buying it for himself but was worried about the cost of repairs so said to us "If you want it, go ahead" as he had lost interest.

'So we put in several bids and won the auction, contacted the owner and shot down to Auckland the next day to pick it up and pay for it.

'On arrival the lady owner handed us a stack of invoices, almost 25mm thick, on everything she had spent on the car since she had owned it. Everything, and I mean everything, has been overhauled or renewed; motor, steering box, brakes, new complete hood, radio aerial. Every W.O.F. it had, new brake pads, wipers and refills, and was serviced with oil and filters. It had only done 5000 kilometres on the motor overhaul.

'On purchase I removed the head which had never been torqued properly. I renewed the head gasket and have spent nothing on the car since.

'It's a 1991 Toyota Celica Convertible, with full leather interior. It's totally unmolested.

'It's got GT-Four [meaning four-wheel-drive] stickers on it which are false. It's a four-wheel-steer, not four-wheel-drive.

'Previous owner had bought another Celica exactly the same to replace this car.'

John Pope's *Corolla*

John Pope has been a lifelong car enthusiast, and remains so to this day. His ownership history is quite eclectic, and spans everything from big Aussie V8 muscle cars, right through to sporty Japanese compacts. During the 1980s and '90s, he owned and raced an XA Ford Falcon GT hardtop, and had a genuine XC Falcon Cobra hardtop to tow it. Today, he races a second-generation Ford Mustang in classic competition.

But back in the early 1980s, he was the proud owner of this stunning green Corolla SR coupe, sporting Cheviot wheels, which were effectively a copy of American Racing Vector-style wheels (as worn on the General Lee in the *Dukes of Hazzard* TV show).

'I used to work at Wrightcars in the early 1980s and they would bring the Toyota rally cars in from overseas for the rally events and then take them away again afterwards.

'I owned this green Corolla coupe at the same time period. I sold my 202 Torana and bought the Corolla. I added the mags and stripes to make it look a bit better. I liked the idea of fitting a twin-cam motor into it but never did. I got promoted to car salesman and given a company car so didn't have it that long.'

At the time, racing and rally cars were a common sight at the company, with several of the crew enjoying a little competition on the weekends. Among them was a second-generation Corolla. In addition, international Toyota rally teams invariably used the company as a base while in New Zealand. Such was the case with a second-generation Celica, which contested the 1980 Motogard rally as part of the World Rally Championship. The factory-supported entry was driven by the French pairing of Patrick Gratian and Jean-Loup Mouledous. They ultimately finished fifteenth, despite hitting something solid during the 1034km multi-day epic.

'Our workshop foreman drove the Corolla rally car. A lot of the guys and girls at Wrightcars were into cars including my boss Bruce Mansell who raced an RX3 rally car.

'There was always interesting stuff coming and going and the hill out the back was good to test the 4-wheel-drives on until someone rolled one down it. We had a 1970 Z28 and an El Camino stored there which I got to drive on occasion.'

The factory-supported Celica of Patrick Gratian and Jean-Loup Mouledous used Wrightcars Toyota as a base for the 1980 Motogard Rally. Its pristine appearance suggests this was before the competition started.

John's foreman at Wrightcars campaigned this Corolla in local rally events. Racing was a popular pastime for many company employees.

Karl Trenberth's Corolla

Off the back of the success enjoyed by its first-generation Corolla, in 1970, Toyota launched the second-generation model of the vehicle that would ultimately become the best-selling vehicle nameplate of all time. The second-generation E20 Corolla was available in a variety of body styles, including four-door sedan, two-door coupe, a van and a station wagon.

The past decade or so has seen popularity in classic station wagons go ballistic, and this enthusiasm applies to Corollas as well. And it's a second-generation Corolla wagon that Karl Trenberth has enjoyed restoring and customizing.

'I was working in a local service station in Christchurch's Stanmore Road at Butler Automart. I saw a nice old lady with a hammer hitting the starter motor inside the engine bay of her Corolla. I went to help her out and asked if she needed a push start. She said, "flipping starter motor had seized and it needed a good whack". I helped her by hitting the starter motor as she turned the key and it burst into life. She asked me if I wanted to buy the car as it had been for sale and no one wanted it. I'd never seen a KE20 wagon before and fell totally in love with it. My sister Anna was looking for a car at the time and this was perfect; one old lady owner with 60,000 kilometres on it. It was a dream come true.

'It was a lovely bile green at this point. She wanted $600 for it. I met her the next day and my sister fronted the money for it. It was then used as a flower delivery car and a camper at festivals for about six years until my sister had trouble getting it a Warrant of Fitness and it was parked up for a while. My sister moved overseas for a number of years and I stored it and kept the registration on hold. It was about 30 minutes away from deregistering one night, but I awoke from sleep and magically remembered to put it on hold for a few more years. Then the Christchurch earthquakes happened. Our garage door fell down on the car in Avonside, the worst hit-area in Christchurch, and damaged the bonnet and roof.

'That was when my old friend Robert Duff, who was the head instructor at SIT Paint n Panel school in Christchurch, came into the frame. He was looking for a nice project for the paint school to do up. He saw potential in the green machine and it was stripped back to bare metal and all the rust and damaged panels were cut and restored. It took about two years and the Krolla [note personalized registration] was reborn.

'It was more beautiful than I could ever imagine. We chose a Toyota colour, 6R4, a beautiful British Racing Green. I took it home and parked it in the garage where it sat untouched for a few years until I met an engineer called Sam Lochrie who asked me why I hadn't started putting it back together. He began to help me with any spare time we had and slowly but surely it began to come together again. We installed a 4K with twin carbs but it had a leak in the block. So we found a worked 5K on the official RWD Toyota page and drove to the West Coast to pick it up. The engine had a lot of work done by a guy called Charlie Bensemann, in Nelson, a renowned rally engine builder. The engine was chemically bathed, bored 020 thou [using later model Series 2 shallow dish pistons], Hypertec coated 30% stronger than standard. The engine was assembled with King main bearings from Israel, NDC big bearing from Japan, Hastings from USA pistons with chrome top rings, Eurotech timing gear from Australia, Permaseal decarb gaskets from New Zealand. It had a Kelford Cam HQR cam grind, Kelford KVS 4K valve springs, Kobi wire [replaces the twin valve springs], CBR machined flywheel @15lbs, Exedy clutch pack from Australia.

'The head has 32cc combustion chamber ported out to 30mm inlet ports and polished, with 10:1 compression, and a Starlet sump. Carburettors are 40mm side-draught Dellortos tuned by Garry Vanderhaven at PAD Racing, making 75 rear wheel horsepower. I still had the 4-speed and standard 4.4 ratio diff. It could barely go 100 km/h, but it got there so fast! It won many drag races but left us deaf after Sam installed a 2-inch custom exhaust. The diff blew one day after some skids at my mate's memorial up the hills so we changed to a u-code Starlet diff and dropped in a K50 5-speed. It now goes 100 mph (160 km/h) but has lost its rapid acceleration with the 3.5 ratio diff. But it's much nicer to drive on road trips now, doing 2500rpm at 100 km/h.

'Auto line shortened my front shocks and added 2-inch blocks in rear with new 14-inch SSR mesh mags to finish it off. NZ Vinyl Trim put new carpet in with a Pioneer head unit and 12-inch sub for ICE. We also installed Mazda RX7 bucket seats with most of interior left standard as it was so mint. I had the car valued by Swann insurance at $20,000 so far as it has now had central locking and key-less entry added with GPS tracking added for security by IS installer service by Angelo. All the locks and key barrel were re-pinned by Ross Gault in Christchurch.

'A new windscreen was installed by JT Auto Glass as the original one was non-safety glass. We have now owned Krolla for 18-plus years and it mainly lives in the garage but gets driven as often as possible. It's been a full-on journey but couldn't have done it without the help of many good friends: Samo Beaton and his amazing brothers for inspiring me, Phil Baines for helping find the hard-to-find pieces, Robert Duff for painting it, Sam Lochrie for the endless nights rebuilding and modifying and tweaking parts to fit, and my fiancée Claire for putting up with me, putting the car first over everything. I hope to pass this car down to my kids one day and they can enjoy it as much as I have.'

MAXXIS
COROLLA

Larry Young's *Hilux*

The fourth-generation Toyota Hilux arrived in 1984, and continued to build on the model's reputation for a tough, durable and reliable work-horse. Being nearly 40 years old now, most have been driven into the ground, and few good examples still remain. But Larry Young is fortunate enough to have purchased this 1985 example when it was barely 10 years old, and has worked hard to keep it mechanically and aesthetically in great shape. It's a true survivor.

'This is my 1985 New Zealand new Toyota Hilux. It's a 2.0-litre petrol engine 4-speed model. It's currently travelled 310,000 kilometres.

'It was originally owned by a large plastics packaging company and used by the engineering department. I purchased it in 1996 to use as a tow vehicle for a race car my uncle and I were building.

'She had already clocked up 200,000kms when I purchased her. She was untidy but sound. I gave the body a good cut and polish and painted the wooden deck.

'As well as towing, other duties included picking up timber, scoria, helped many people move house and occasional trips to the dump to get rid of garden waste etc.

'Helping out at driver training days it was my job to lay out the cones to show where the apexes were. While taking the opportunity to do 2–3 laps I managed to get close to 130 km/h out of her down the back straight of Pukekohe. Once I used her for an uphill gymkhana against a Falcon ute. Unfortunately the bad turning circle hindered my loop at the top cone, but off the mark we were tied.

'I fitted polished Trident wheels soon after I owned the vehicle. Last year I refurbished the rear frame to get rid of some rust and used a high-quality hardwood for the tray. It's too good to throw scoria in there now!

'Having owned her for over twenty two years, she has given me no issues whatsoever. Parts are becoming scarce but I'll keep her running for as long as I can.'

Mathew Robertson's Corona

Toyota released the Corona in the Japanese domestic market in 1957, and while the second-generation model was exported to the United States in small numbers, it wasn't until the arrival of the third-generation variant in late 1964 that Corona became a genuine international car.

It was the third-generation Corona that first came to New Zealand, and by the time the fifth-generation model was released in late 1973, it was a common sight on Kiwi roads.

These days, of course, early Coronas such as the one owned by Mathew Robertson are a rare sight. Mathew's example makes full use of the model's great styling, which is improved upon with lowered suspension and vintage Japanese custom wheels.

'This is my 1978 Toyota Corona RT104.

'When I was younger I had owned a few KE Corollas and liked the look of the old Coronas. So one day I decided to buy one.

'It didn't look anything like it does today. About 15 patches were cut out and new steel welded in to replace the rotten steel.

'The motor is an 18RG out of a Celica which was imported into New Zealand from Japan as a guaranteed runner with low mileage back in the early 1980s to be put into concept cars by a man who lives in Palmerston North. It was always turned over by hand but this particular 18RG he'd never got running until I bought it off his mate Mike Jenkins Trading in Palmerston North and put it into the Corona. The engine and gearbox have done bugger-all kilometres since new in the 1970s.

'The 18RG came out in certain GT models of this Corona so the conversion was pretty much bolt-in.

'The wheels are SS05 Superstars from Japan which I had refurbished here in New Zealand.'

Michael Wilson's *Celica*

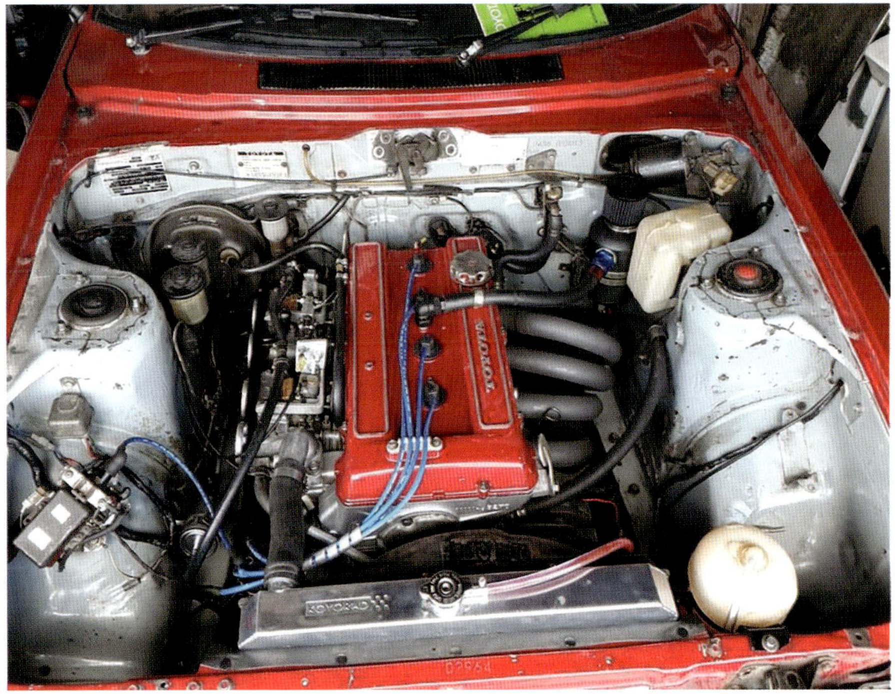

Toyota launched the first-generation Celica in 1970, aiming it squarely at the North American market. And to help establish it with US buyers, stylists borrowed several design cues from the first-generation Ford Mustang, which spawned the pony car market.

First-generation Celicas were available as either a two-door hardtop or a three-door liftback/fastback, and when comparing it to the Mustang of the era, the similarities are clear to see.

But the Celica wasn't just a North American Toyota. It was a global Toyota, for it sold in countless markets, including here in New Zealand.

From the outset, Toyota chose the Celica as its weapon of choice in the highly competitive Japan Touring Car Championship, where it battled the Nissan Skyline and Mazda RX2 and RX3. And here in New Zealand, it was also a popular competition model for a variety of disciplines, from circuit racing, rallying, and even the Targa. Michael Wilson owns this superb example with multiple New Zealand Targa events under its belt.

'The Targa Celica started rally life with Alan Moore in 1999. I believe it's a car he owned a lot longer than that and brought here from Australia with him, which he had sold but managed to buy back for the Targa.

'Alan used it in Targa and I service crewed for him. I then co-drove in it for the 2004 Targa 10th anniversary. It was then raced at the classic meets at Manfeild, and finally retired in 2008.

'I purchased it from Alan a few years ago and I'm slowly tidying it up with the intention of using it for classic racing again, trying to keep it resembling what it looked like in the Targa.'

Mikayla Nelson's Starlet

With each new model generation, cars tend to get bigger and bigger. Eventually, they drift so far from their original purpose, the manufacturer ultimately creates an all-new model to fill the void they've left behind.

When Toyota first launched the Starlet in 1973, its intention was for the sub-compact market. New Zealand didn't receive the first-generation model, but we did get the second-generation, released in 1978. Even by the late 1970s, this was a small car. But it was only 300mm shorter than the first-generation Corona, which by the late 1970s, was itself carrying an extra 300mm of girth over its original 1957 sibling.

The Starlet, however, retained its diminutive dimensions, and four generations later when the model was eventually culled in 1999, it had barely grown.

First- and second-generation Starlets became hugely popular in rallying during the 1980s, thanks to their short wheelbase and rear-wheel-drive. Today, good examples are incredibly hard to find, but they're still quite affordable, offering perhaps the most fun for the money of any classic car. And furthermore, they respond well to subtle modifications, as demonstrated by Mikayla Nelson's second-generation example.

'This is my 1984 KP61 Toyota Starlet. I have owned the car since late 2017, and in that time I have chipped away bringing the car back to its former glory.

'You're probably wondering why I say former glory, but the car is still quite modified. About five years ago the Starlet was still fairly standard, but my partner's brother was asked by the owner at the time if he would be able give it a new lease of life. This included a full panel and paint in the blue hue you see.

'It's got KE30 chrome bumpers which have been shortened to keep them snug to body. The suspension has been lowered some more, with the fitment of 15x8 inch wheels. To get these to fit, some modifications to the rear arches were needed, but we wanted to keep the original look of the arch while giving the wheel more clearance.

'Once the car was all finished the owner used it for about a year, then it was sold on. It went to the South Island and had a bit of a hard time down there. About four years on I was looking for a new car. I took a bit of a liking to Starlets and my partner jokingly said "you should buy this". And so I did.

'We brought it back to Wellington. It was looking a bit sad for itself. Rust was showing up in a few places, bumpers dented, grille all out of alignment, and dents all through the roof. But I wasn't thrown off by all this. I was determined to get it looking as good as it was those five years ago. The next day after bringing it back, we stripped the front-end apart, and put it all back together, making sure everything lined up good. We straightened out the bumpers along with a few other things. Even after just one day it looked like a whole new car.

'A few months on, it needed a WOF but subsequently failed on some rust. So the decision was made to pull it off the road, fix all the rust, panel beat it, and paint it in bright blue again but with a slight difference. At this time, the interior was stripped out, new carpet installed, and new stereo system along with a custom speaker box in the boot.

'The wheel centres are now painted, along with a subtle white pin-stripe along the centre of the car, giving this Starlet what I think is a very clean look and a new lease of life, which I am glad to be a part of.'

Nifty Customs' Celsior

This beautiful creation is the latest promotional vehicle for Nifty Customs. It's a third-generation Toyota Celsior.

Toyota launched its flagship Lexus LS brand in 1989, to tackle the international luxury vehicle market largely dominated by German manufacturers. Until the arrival of the fourth-generation Lexus in 2006, Toyota marketed both the Lexus LS 400 and LS 430 in Japan as the Celsior, to side-step complications with its existing four dealership networks. Hence, with the strong used Japanese import market, the Celsior has also reached Kiwi shores.

The third-generation Celsior utilized Toyota's impressive 4292cc 3UZ-FE V8, which produced 216–224 kW (290–300 horsepower), depending on its application and year. The 3UZ-FE featured an alloy block, topped with quad-cam cylinder heads and 24 valves. Such was its broad usability, the V8 was also used in Toyota's Super GT racing in 4.5-litre configuration, and even as a 5.0-litre in the American Grand-Am series.

Despite the Celsior being a vehicle dripping with incredible Toyota technology, it isn't a model widely adopted by car enthusiasts for modifying. Many tend to gravitate towards the smaller, more nimble and naturally sporting models. But the big Celsior has a champion in Vadim, owner of Nifty Customs, who has applied his creativity and craftsmanship to high-end Japanese luxury cars.

Like any good custom car, the bodywork modifications featured on Vadim's Celsior are subtle enough that they improve both the look and quality of it, but at the same time, appear natural in their surroundings.

'This is my first Celsior, but I have owned a Toyota Crown in the past. That one was also done in traditional VIP style.

'The bodywork on this car consists of a custom modified K-Break bodykit, custom front fibreglass vented guards, rear sports custom modified Mercedes E63 AMG diffuser moulded into a modified Aimgain bodykit, and AMG exhaust tips. The boot lid is finished off with a K-Break lip and a subtle roof spoiler.

'The rear quarters have been cut and Ulterior Motive flares have been welded on and moulded to the body, essentially making this a widebody.

'This car came with factory air-bag suspension that I have modified to operate like an aftermarket one. But this has been changed to UAS air-bags, XYZ coil-over struts and Airlift 3P management. The new setup allows for the car to fully "air out" and the subframe sits on the floor.

'The interior features front seats which have been re-foamed using Nissan 350Z foam and leather creating almost a bucket seat effect.

'The rear seating has been converted to a two-seater by deleting the middle seat, installing electric reclining frames and motors and custom-made middle console.

The Celsior, as originally purchased by Vadim. Its transformation would be immense.

'My plans for the Celsior are never ending, with the car and its list of parts constantly growing.

'The next stage will be installing Porsche Panamera 6-pot Brembo brakes with 390mm rotors all-round.

'I'll further modify the front fibreglass fenders to match the rear guards. I'll finish off the interior and colour change to suit 350Z seats. There are also many areas on the car that still need attention and some minimal changes that will happen. It'll also receive a colour change from pearl white to something different, which at the moment is a secret.'

As far as promotional vehicles go, Vadim's Celsior is certainly one of the very best, showcasing the services Nifty Customs offer. It's a company that is growing at pace, from humble beginnings, in the most unlikely of times. 'Nifty Customs started up during the first lockdown in 2020. It started as just a part-time gig in my shed at home. But once I figured out there was a big demand for my services, I ramped it up and opened a proper workshop in August 2020. I specialize in fibreglass repairs, plastic welding, fibreglass bodykit fitting services and widebody conversions. Next step for Nifty Customs is a much larger workshop with paint booth and a full-time painter.'

Paul Hollingsworth's Corolla

Although the first-generation E10 Corolla was sold briefly in New Zealand in the late 1960s, the superb-looking two-door Sprinter coupe wasn't an option for Kiwi buyers. But that was all put right with the introduction of the second-generation model in 1970, which offered a stylish coupe for the first time.

The Corolla coupe was the hot ticket among New Zealand car buyers in the 1970s, and greatly aided Toyota New Zealand in raising its profile. These second-generation models are rarities on Kiwi roads these days, and even more so the desirable coupe, which is understandably sought after among classic Toyota enthusiasts.

Paul Hollingsworth, a lifelong Toyota enthusiast, owns this great example mixing the classic second-gen coupe styling combined with more modern technology, and is a true tarmac-shredder.

'I've been a Toyota Corolla owner since I was fourteen.

'My first Corolla was a Citric Acid Green 1978 KE35 SR Coupe which started my love for Corollas. I owned that car for around four years and then sold it on and bought myself a very standard KE36 by following an old grandma home. But when the standard wagon started to get a little boring, I sold it and started building a faster model.

'I purchased a bare 1973 KE20 shell and a Mitsubishi 4G63 from an Evo 6, and proceeded to engineer it all together with a Supra gearbox and Hilux diff. When a friend decided to sell his KE25, I purchased the rolling body and swapped all my running gear over.

'A few years later she was on the road and I've been improving her ever since. Since that project was done, I ended up buying a KE26 Corolla that had been off the road for nearly twenty years and had it sand blasted and I'm in the process of a full restoration. So I almost have the full collection.'

NVMY20

MITSUBISHI

NVMY20

NVMY20

The Radisich Trueno

Paul Radisich is one of New Zealand's most celebrated, most successful racing drivers. Back in the 1990s, he achieved something very few New Zealand drivers could manage: a fully paid international racing career.

Radisich enjoyed a rapid rise through the single-seater ranks very early in his career. In just his second New Zealand Grand Prix at Pukekohe in January 1983, he finished second to David Oxton, in a Ralt RT4. He also contested the 1983 Australian Drivers' Championship where he impressed with a number of good results.

Back in New Zealand Paul finished first, fourth, third, and sixth in the four-round New Zealand International Formula Pacific Championship, before contesting the first five rounds of the highly competitive British Formula 3 Championship. He took pole position in just his second race, at Thruxton, and demonstrated his talent with a couple of top five finishes in his handful of appearances.

After a quiet 1985, Radisich returned to British Formula 3 in 1986, driving a Murray Taylor Racing Ralt RT30 alongside Damon Hill. He and Hill were evenly matched; Radisich's best finish was third at Snetterton, one place behind his team-mate, who would go on to win the World Championship in 1996.

1987 saw Radisich move to the US where he contested Formula Super Vee, and Indy Lights. Meanwhile, he was a regular in the New Zealand Formula Pacific Championship, winning the title in 1988. In addition, Radisich made infrequent appearances in touring cars, most notably at Bathurst in 1989 and in 1990. He and team-mate Jeff Allam finished the 1990 race second.

Radisich continued to aim for Formula 1, but this never eventuated. Where he did land, however, after years of struggling, was the hotly contested British Touring Car Championship. In 1991, the series switched to an exciting new formula called Super Tourers, with a maximum 2000cc engine size. The cars were fast and spectacular, and manufacturers waded in, spending millions in an effort to have their products win in front of record spectator attendances and a worldwide television audience.

For the 1993 BTCC, the hugely successful Andy Rouse, who'd won the championship four times, switched from running factory-backed Toyotas to factory-backed Fords, debuting the all-new Mondeo V6 in the series. His new team-mate was Paul Radisich.

Despite having missed the first eight races, Radisich finished an impressive third in the championship behind the factory BMWs of Joachim Winkelhock and Steve Soper. Moreover, he ended the season with nearly three times the amount of points Rouse accumulated. Suddenly, the world knew who Paul Radisich was. And more was to come.

In October 1993, a massive Super Touring competition was contested at Monza, Italy, called the FIA Touring Car World Cup. In all, 45 drivers were entered, representing all the countries using the Super Touring regulations.

There were two heats, and the winner would be based on accumulated points. Paul Radisich qualified fastest, and went on to win both heats, to be awarded the FIA Touring Car World Cup Champion. Furthermore, there were accumulated National Standings, based on country of origin for the drivers. Radisich finished third, despite being the only Kiwi on the grid.

Radisich had a strong season in 1994, winning two races and taking seven runner-up positions to finish the championship third once more. He then rounded out the year by again winning the FIA Touring Car World Cup, held at Donington Park.

After four less successful seasons in the UK, with Ford and Peugeot, Radisich left the BTCC in 1999 and took up an offer from Dick Johnson Racing to drive a Falcon XR8 in Australian V8 Supercars. He'd already shown his adaptability in being able to jump from a front-wheel-drive 2-litre BTCC car to a rear-wheel-drive 5-litre V8 by performing impressively at the 1996 Wellington Street Race.

At DJR, Radisich drove alongside Johnson, winner of two Bathurst 1000s, and five Australian Touring Car Championships, and quickly became the team's lead driver. He won several races, and looked to be heading for a Bathurst victory in 1999 before hitting a lapped car and puncturing a tyre while in the lead. He finished second at Bathurst in 2000. The following few seasons provided few good results and, in 2006, Radisich's season came to a premature end at Bathurst when his Team Kiwi Racing Commodore hit the wall exiting the Chase, and rolled. He sustained a broken ankle and sternum. By 2008, Radisich was winding down his full-time commitments; during practice at Bathurst he crashed the Toll-HSV Commodore he was sharing with Rick Kelly, when the throttle jammed open over the high-speed McPhillamy Park, and slammed the concrete barrier at the exit of the turn. He suffered multiple injuries, which brought an end to his professional career.

Paul Radisich has had a highly successful career, and is now a much sought-after driver in historic racing events, particularly in Europe, where he is still greatly revered.

But long before the international fame, and factory drives, Paul Radisich shared a Toyota Trueno Sprint with his father Frank, in the 1981 Benson & Hedges endurance series for production cars.

The B&H series was contested late in 1981. At the time, Radisich junior was just 19 years of age, with an exciting career ahead of him. Radisich senior, however, was bringing his racing career to a close. Frank Radisich, much like his son would prove to be, was highly adaptable, and could jump into and race just about any type of car, from single seaters to saloons. In the 1970s he switched back to saloon cars, and much of his racing was in the popular production class. Unlike his son, Frank Radisich never raced as a professional. He ran a garage in Auckland called Henderson Central Motors.

New Zealand held something of a fascination with production car racing, going back to the 1960s. Much of this stemmed from the thriving local vehicle assembly industry. During the 1960s and 1970s, an annual production car endurance race, with a distance of 500 miles (804 kilometres) was contested at Pukekohe. Only New Zealand assembled cars were eligible. The race was sponsored by Benson & Hedges cigarettes, and was named the B&H 500. In 1978, the race distance changed to a fixed time of five and a half hours. From 1981, the B&H event evolved into a three-race endurance series, contested in the North Island at Pukekohe, Bay Park and Manfeild. The regulations still required cars be New Zealand assembled, and there were four classes, based on engine size, with the largest being Over 2000cc. However, recognizing that the B&H 500 races had been mostly dominated by a single marque, the endurance series rules allowed for imported cars, of up to 2000cc. But these cars were placed in the Over 2000cc class, against the V8 Holden Commodores and Ford Fairmonts.

These were hotly contested events, and featured some of New Zealand's best racing drivers. 1967 World Formula 1 Champion, Denny Hulme, shared a Commodore with two-time New Zealand Saloon Car Champion and 1974 Bathurst podium finisher Rod Coppins, while multiple B&H 500 race winners Leo Leonard and Gary Sprague were entered in a V8 Fairmont. Open wheeler hot-shoes David Oxton and Garry Pedersen shared a Commodore. Nick Begovic got Dick Johnson across from Australia to co-drive his Fairmont.

For the Radisich clan, the B&H series was a family affair, as it had been the year prior. In 1980, Frank shared a Mini with his wife Robyn, while their sons Paul and Chris drove a Mazda 323. For 1981, Robyn drove a 1275 Mini with Chris, while Frank and Paul shared a Toyota Trueno Sprint. The Trueno was owned and prepared by Frank. Because it was not New Zealand assembled, the Trueno was placed in the Over 2000cc class.

The opening race was at Bay Park, and Frank and Paul acquitted themselves well to finish fifth, behind the winning Commodore of Oxton/Pedersen, Rod Collingwood/Lin Neilson in a Volkswagen Golf, Leonard/Sprague, and Coppins/Hulme.

The next race was held at Manfeild, and here the young Radisich really hit his stride. He got involved in a three-way tussle for the lead with the Leonard/Sprague Fairmont, and the Toyota Celica GT of Kent Baigent/Graeme Bowkett, which was also entered as an import. Radisich got faster as the race progressed, and was mounting a challenge on the Fairmont for the lead. But just as he was about to pounce, he suddenly slowed. 'We ran out of fuel,' recalls Paul Radisich today. 'We fielded two cars in the race [the other being the Mini] and the wrong fuel went into the wrong cars at the fuel stop! We were waiting for a miracle to happen but it didn't. There were no pace cars in those days.' In the end, it was the Baigent/Bowkett Celica which took victory.

The Trueno was a handsome-looking car, and a rarity on New Zealand shores, even in the early 1980s. Does it still exist?

Indeed, the Radisich Trueno was running in fifth place with 55 minutes to run. They stopped to add oil, but only got as far as the back straight before the car was retired. However, the 19 year old had impressed a lot of people with his speed and race craft. And this would ultimately pay dividends.

Paul Radisich's drives in the family Trueno in 1981 are largely overshadowed by the international successes he enjoyed later in his career, but his drive in the Trueno got him noticed. And that kick-started his racing career. 'The Toyota was the second saloon car I drove,' explains Paul today. 'The first was with my brother Chris in a Mazda 323 the year before in the B&H race at Pukekohe. But the Toyota Trueno launched my single-seater career, firstly with [Formula Pacific team owner] Dave McMillan in the New Zealand International Grand Prix, and then with our own Ralt RT4. Garry Pedersen [driver and team owner] convinced Dave McMillan to give me a drive in his RT1. That's how I got a start in single seaters.'

And the rest, as they say, is history.

Rhys and Nick's Corolla

When Japanese car manufacturers began establishing themselves in New Zealand during the late 1960s, they used rallying competition as a form of marketing, to help promote the strength, reliability, and sporting attributes of their vehicles. This continued into the 1970s, and rallying quickly became a popular domain for a variety of Japanese brands.

Toyota entered rallying not long after establishing itself on Kiwi soil, initially with the Corona, and then the Corolla. Both models enjoyed plenty of success.

These days, early Corollas remain popular rally cars among Kiwi enthusiasts, such as this four-door second-generation model campaigned by good friends Rhys and Nick.

After deciding to dabble in some rallying from the driver's and co-driver's seat, Nick and Rhys found a 1975 Toyota Corolla that had been run by Shane Cameron in Canterbury. As it was a classic rally car built to the specifications of the period it was from, Motorsport New Zealand granted a 'Certificate of Description' (COD), deeming the car to be eligible for classic motorsport events. The aim of rallying a classic car appealed to the new owners after being involved in servicing for a number of top-performing 4WD and 2WD open rally cars where all the hard work can be wasted due to the failure of a single part; a recipe the pair saw continuously strip the fun component out of rallying.

The car was shipped to Palmerston North where a strip-down and rebuild commenced to tidy up the overall appearance of the car and the blown engine that was a result of Tauranga Rally 2014. Over the course of the following two years a few rallies and gravel sprints were completed with the car performing well. Various parts were upgraded along the way with the basics of the engine consisting of a 4K 1300cc block, Arias pistons, lightened and polished crank, ported and skimmed head, race camshafts and 40mm DHLA Dellorto carburettors producing a static compression ratio of 13.6:1 and just

shy of 90 horsepower (over double the output from factory).

An AE86-sourced T-series 5-speed gearbox with a TRD close-ratio gearset was added to get power to the rear wheels where another AE86-sourced T-series unit gets power to the ground with the assistance of a TRD 1.5-way limited slip differential and Bilstein shocks. This combination of engine and drivetrain certainly doesn't aid grocery-getting performance, being almost undriveable below 3500rpm due to the large cam, carburettors and an extremely tall first gear, but once moving on twisty gravel stages the engine reliably sings between 4000 and 7500rpm with perfectly matched gears to utilise the high RPM power of the engine.

Manawatu-based Day Breaker Rally was the first successful competitive rally for the duo and to keep the fun component as the top priority the owners always swapped driver and co-driving positions between each rally stage. The event didn't go completely to plan after the alternator wiring caught fire during stage 1. The car was fixed on the side of the road using some wires borrowed from the farmer and the pair re-joined to be competitive to the end of the rally. But most importantly a massive amount of fun was had!

After competing in local Dannevirke car club gravel sprints in 2015, the goal was set to complete the Tauranga Rally again in 2016 after the disappointment of 2014. Just 6km into the second stage the car rolled off the road causing extensive damage to the roof, panels and roll cage.

The panels and roof were repaired and chassis seam welded to improve rigidity. Being an engineer by trade Nick fitted an extensive 8-point roll cage, custom stainless exhaust, rebuilt every bracket and part; the entire car was restored to an immaculate factory-fresh finish. This process was the result of thousands of hours of hard work and took over 18 months to complete. Future events are planned to use the improved car and make rallying fun again!

Team Castrol
RT1778
40

Richard's
Carina Super Tourer

We quite rightly think of Greg Murphy as a faithful Holden driver. And certainly, his decorated V8 Supercar career was spent racing General Motors products. But there were other brands peppered throughout that also played pivotal roles, and one of them was Toyota.

Although a devout open wheeler driver early in his career, Murph ultimately switched to tin tops, and went on to enjoy a hugely successful V8 Supercars career, notching up 28 race victories and 81 podiums. His greatest racing achievements were surely his four Bathurst 1000 victories, in 1996, 1999, 2003 and 2004. However, it wasn't in a Holden Commodore that he made his Bathurst debut. Instead, it was a Toyota Carina.

Back in the 1990s and early 2000s, the 2-Litre Super Touring formula looked to be taking over the world. The concept was brilliant. It catered to mass-produced, four-seat, four-door family saloons, with a maximum engine size of 2000cc. But although a strictly controlled category, there were some incredible-looking cars and epic racing.

Super Touring regulations demanded a largely stock bodyshell, with only factory aero aids, although this ultimately changed to prevent manufacturers getting too clever and building homologation specials draped in aerodynamic garnishing. But what really made the Super Tourers stand out was their massive wheels, which measured an astonishing 18 inches in diameter initially, before moving to even larger 19 inches.

Although they looked like regular family saloons, albeit with huge wheels tucked way up into the bodywork, they were anything but. These were ground-up, purpose-built racing machines, and were completely customized and dripping with Formula 1 quality detail and fabrication.

Manufacturers waded into supporting 2-Litre Super Touring virtually from the outset, and it was the British Touring Car Championship (BTCC), where the formula got its start, that the factory teams were most abundant. Indeed, throughout the 1990s, the BTCC boasted manufacturer teams from BMW, Ford, Audi, Nissan, Renault, Volvo, Peugeot, Mazda, Vauxhall, Honda, Alfa Romeo, and, of course, Toyota, and they pumped millions of dollars (or pounds really!) into the series.

The 2-Litre Super Touring formula debuted in the BTCC in 1991. And from there it quickly spread. In 1993, it was adopted for the French, Italian, and Portuguese Touring Car Championships, while also being introduced in Australia. 1993 also saw the first World Touring Car Cup, an event in which Super Touring teams from around Europe and Australasia all converged for a single contest.

In 1994, 2-Litre Super Touring became the formula of choice for the Japanese, German, Belgian, and South African Touring Car Championships, in addition to the reintroduced Asia-Pacific Touring Car Championship. For us here in New Zealand, 2-Litre Super Touring was chosen exclusively for the Nissan Mobil Series; the Wellington Street Race event and its Pukekohe partner. More countries adopted the formula later.

But somehow, it was the BTCC, where it all began, that remained a stronghold, and its incredible quick-fire live television feed, complemented by the screaming tones of famed commentator Murray Walker, made it an absolute sensation.

Factory Toyota driver Will Hoy almost won the 1992 BTCC, aboard an Andy Rouse Racing Carina. The Carina was the British variant of the Corona, and was the same in all respects, other than badging. Toyota was quick to involve itself in the 2-Litre Super Touring formula, and Hoy was one of three drivers capable of winning the title going into the final race at Silverstone, alongside BMW driver Tim Harvey and Vauxhall driver John Cleland. Indeed, this race was famous for the dramatic clash between Cleland and Harvey's BMW team-mate Steve Soper, which witnessed the demise of both drivers. Hoy's Carina faded during the race with mechanical troubles, and Harvey won the title.

For the 1993 season, the new tenth-generation T190 Corona/

Carina E made its BTCC debut. Four factory-supported cars were entered, all built by TOMS (Tachi Oiwa Motor Sports), which ran a pair of cars itself for Will Hoy and Julian Bailey, while the Park Lane Toyota Junior Team ran another two TOMS-built cars for James Kaye and Bobbie Verdon-Roe. The Park Lane team was run by Formula 1 driver Derek Warwick, as well as Roland Dane, before the pair ultimately partnered with Ian Harrison to establish Triple 8 Race Engineering in 1996. Some years later, Dane would expand on the Triple 8 programmes in Australian V8 Supercars.

One of the Park Lane Toyota Junior Team cars was chassis 015-005, and was driven by team leader Kaye in all but the final two races, when it was driven by Verdon-Roe. It ran the white and green colours of sponsor Enny. In what was an intensely competitive BTCC season, Kaye scored a best result of ninth in Round 3 at Snetterton. Verdon-Roe in the sister car did slightly better, notching a best result of sixth in Round 3 at Snetterton, seventh in Round 5 at Oulton Park, and eighth in Heat 1, Round 6 at Brands Hatch.

Of the two TOMS factory team cars, Bailey scored victory in Heat 2, Round 9 at Knockhill, while Hoy's best result was second in Round 7 at Pembrey.

The second half of the 1993 BTCC saw the arrival of the factory Ford Mondeo team, with two cars for team owner Andy Rouse and new recruit, Paul Radisich. New Zealander Radisich was hugely impressive, winning three of the last five races, and scoring runner-up spots in the other two.

The 1993 TOMS-built Carinas were pure works of art, at the heart of which was a heavily modified Toyota 3S-GE 4-cylinder 16-valve motor producing around 300 horsepower at 8500rpm, and around 220 lb/ft. They ran a Zytek multipoint engine management system, with programmable fuel-injection, and a dry-sump with the oil reservoir mounted in the engine bay.

Backing the super trick motor was an Xtrac 6-speed sequential gearbox with quick-change dog engagement gears, and an AP Racing clutch. The cars rode on Toyota Super Strut front suspension with MacPherson struts in the rear, with massive 14-inch-diameter Brembo ventilated discs doing the stopping. Wheels were 18-inch-diameter magnesium OZ. Instrumentation was TMS (Toyota Management System) data-logging computer with programmable data display.

Thirty years ago, this was verging on Formula 1 technology. Little wonder the teams ran such massive budgets.

For the 1994 BTCC, TOMS built a new fleet of Carinas, and chassis 015-005 was sold to New Zealand businessman Bruce Miles to contest the new Australian Class II Championship, as the popular 2-Litre Super Tourer formula spread its way down-under.

Australia, at this time, had become fiercely divided over the future direction of touring car racing. The 1993 series saw a three-class system, which included the new 5-Litre Touring Cars, 2-Litre Super Touring, and existing outgoing Group A cars. By 1994, the 5-Litre Touring Cars and 2-Litre Super Tourers ran their own championships, and their own race schedules. The 5-Litre Touring Cars would eventually be rebranded as V8 Supercars. The 2-Litre Super Touring series would be rebranded the Australian Manufacturers' Championship.

Run by Campbell Little (who would go on to become one of the most decorated V8 Supercar race engineers), Carina 105-005 was driven by both James Kaye and the virtually unknown Kiwi, Greg Murphy, in selected rounds of the 1994 Australian Manufacturers' Championship. Kaye, who was also contesting the BTCC, competed at Winton and Lakeside Raceway, scoring a second place finish at each event. Murphy, meanwhile, finished third in Heat 1 at Phillip Island, fourth in Heat 1 at Oran Park, and third in Heat 2 at the same event. The championship was largely dominated by the factory BMWs driven by Tony Longhurst and Paul Morris.

The 1994 Bathurst 1000 featured a combined grid of 5-Litre Touring Cars and 2-Litre Super Touring, and it was here Kaye and Murphy teamed up to share 015-005 in what would be Murphy's Bathurst debut. The 5-Litre Touring Cars would contest Class A, the 2-Litre Super Touring cars Class B. Naturally, the V8s were much faster around the Mount Panorama circuit. Of the 11 Class B entries, Kaye and Murphy qualified an impressive third behind the two top factory BMWs, and ahead of the third factory BMW, which was driven by Kiwis Craig Baird and Brett Riley. They lined up 33rd outright from a total 47 cars.

Come the race, however, various delays shuffled them down the order, to eventually be classified 23rd outright, and fifth in Class B.

As for the Toyota Carina, lack of sponsorship meant it didn't race in 1995, and indeed, didn't turn a wheel in anger until 1999, by which time it had a new owner. Its only outing was at Bathurst, for the Bob Jane T-Marts Super Touring 500. The Super Touring cars were by now contesting a completely separate Bathurst enduro from the V8 Supercars. The Super Touring race was held in the traditional early October slot, while the V8 Supercars had their competition one month later.

In the 1999 event, the Carina was entered for Clayton Haynes and Carlos Rolfo, but they failed to start. It was mothballed again until 2002, where it was raced by Rolfo and Claude Elias in the Australian Super Touring Championship (formerly Australian Manufacturers' Championship). Elias took a pair of seventh place finishes in both heats at Oran Park, while Rolfo finished fifth in Heat 1 at Mallala.

And with that, 015-005 was effectively retired, and received very little use until purchased by Richard McCarthy in late 2017. Richard has since fully restored the Carina, bringing it back to its 1993 Enny livery as raced by James Kaye in the BTCC. Richard's brother Paul also owns a former factory 2-Litre Super Touring car, a super-rare Mazda Xedos 6, which actually competed against the TOMS-built Carina in the 1993 BTCC, driven by Patrick Watts.

Richard and Paul run their incredible 2-Litre Super Touring cars with the Heritage Touring Car group, which also caters to genuine Group A, Group 2 and Group 5 cars with period race history.

As for the 2-Litre Super Touring formula, its success ultimately paved the way to its demise. With so many manufacturers involved, the pressure and expectation to win spiralled, along with the costs to do so. Funnelling millions into a racing series and not getting the desired returns eventually had the manufacturers pulling out once more, and by the year 2002, it was all over. The BTCC switched to a new formula that focused on controlling costs, and most other championships throughout Europe did likewise. But the epic, extraordinary cars that were built for the 2-Litre Super Touring category are now in high demand among collectors, and make regular appearances at many of the top historic racing events around the world.

demon tweeks
Kaye
10
Auto Trader
Bilsom
THE food WEIGHOUSE
TRUST
Castrol
YOKOHAMA
TOYOTA
TOYOTA junior team
PARK LANE

Kaye
10
TRUST
Bilsom
Castrol
TOYOTA
YOKOHAMA
NGK
MAY S
MOTORS
TRANSPORT LTD
momo
WILLANS
PARK LANE
Auto Trader

RACE | RALLY ACCESSORIES
Kaye
10
TRUST
Bilsom
Castrol
MAY S
MOTORS
momo
WILLANS
PARK LANE
Auto Trader

Richard Opie's Starlet

Chances are, you know who Richard Opie is, even if you've never met him. His stunning photography graces the pages of several car magazines, most notably, *New Zealand Performance Car*.

Richard's little 1981 P60 Toyota Starlet is as sharp as his photos. It's been a labour of love, one in which the transformation spanned more than a decade. And the results are truly stunning.

'I bought this car mid-2005. At the time I had an Alfa Romeo Alfasud Ti that was my pride and joy; restored as a teenager. I bought the KP61 (the 1300cc model) purely as daily transport to work and university, to preserve the Alfa. It was an original, unmodified car but pretty banged up. It had a career as a courtesy car for a garage in Addington, Christchurch.

'This Starlet is a little bit unusual, as it is a Japan market car imported here in 1987. They differ a little from the Kiwi new cars; mainly just wee trinkets, oddball design features and much nicer trim generally. It's a Toyota Starlet XL Lisse. This was a model aimed at women in its home market. From factory it came fitted with a 4K 1300cc engine, which New Zealand models didn't have at this point.

'My Toyota ownership is only KP Starlets! I've had six of them in varying states. The bug was kicked off with the purchase of a $100 car that I crudely patched the rust on, then got it WOF'd and blazed around the mean streets of Christchurch in. It pretty much sold me on how fun the little RWD [rear wheel drive] chassis could be, even though it had a rattly old 2K (998cc) in it.

'The build on this [our feature car] took me 11 years. It came off the road for a 'clean and tidy', then things ballooned. It was initially going to be a warmed up 5K with bike carbs, no interior, no flash suspension and brakes. Scope creep took hold along the way, as you see what others can achieve and continually tell yourself, "well, it doesn't seem like THAT much work to go the extra mile on that." Case in point is the interior - I never intended to completely retrim, just repair and trim the worn front seats. But I figured if I was spending the time and resources on that, then I may as well do this.... the rest is history.

'The engine is a 16v 4A-GE, with 3-rib bottom end, balanced rotating assembly, TRD forged high compression pistons, late smallport head with porting, 1mm oversized valves, Kelford 193B camshaft, Supertech valvetrain et cetera. Quad throttle bodies off a 4A-GE 20V 'silvertop', high-rise header by Brendan Duncker, and a stainless 2.5-inch exhaust. Engine management is Link G4+.

'The gearbox is an AE86 T50 5-speed, with Toda 3.6kg flywheel, Exedy heavy duty clutch. Diff is AE86 Kouki (facelift) T-code 6.7-inch, shortened to fit KP width. It uses a TRD 2-way clutch LSD, with T3 rose-jointed 4-link arms.

'Front suspension is Fortune Auto coil-overs fabricated to suit KP spindles and adjustable camber plates. Rear suspension is Bilstein with adjustable height platforms on the diff.

'Brakes are Wilwood Dynalite 4-pot front, and AE86 disc rear. It's also using a Wilwood bias adjustable pedal box and master cylinders.

'I change wheels like I change undies! Restoring vintage Japanese wheels is a hobby of mine, but the favourites are my Hayashi Street ST in 13 x 7-inch with +7 offset front, and 13 x 7 –6 offset rear with Yokohama A539 tyres.

'The interior features Recaro LX "fishnet" seats (classic fitment to these by boy racers in the "golden era" of the Kiwi street scene), and new door cards as the old ones resembled soggy Weet-Bix. The new cards also leave the painted door tops exposed, versus the originals that covered them up (it was a higher trim level). But I prefer having the painted surfaces exposed. It's all trimmed in a maroon vinyl with wool houndstooth inserts. Maroon, to match the factory maroon plastics, and the houndstooth as a nod to the period Starlet fabrics

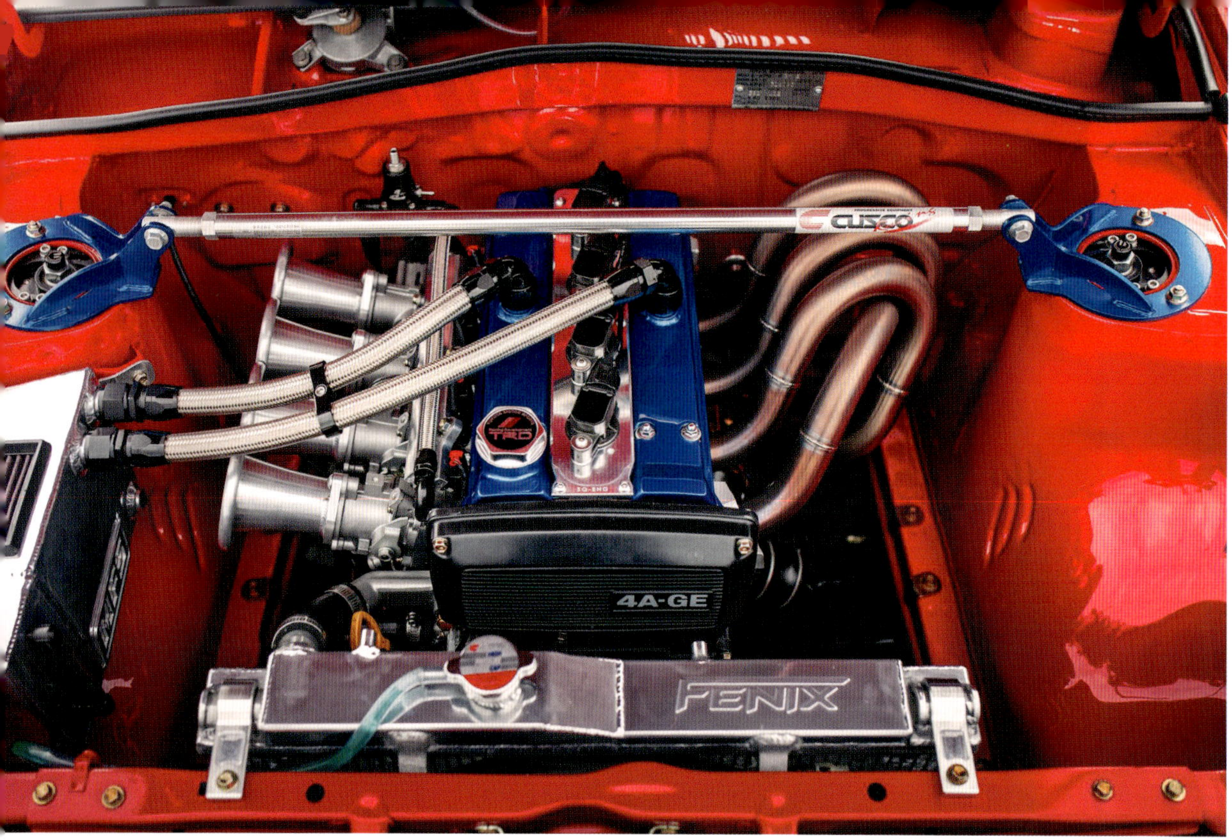

used, and also the European interiors of the '70s that I really adore. It's also got Stack instruments throughout, something I have always wanted ever since watching motorsport coverage as a kid in the '80s. The Nardi Classic steering wheel is almost mandatory for old Japanese hardware, I think? It also has period-correct audio, in the form of early '80s Japanese-market Pioneer component stereo and speakers.

'The paint and bodywork is simply factory red, with no crazy mods. I wanted to retain the factory lines. To me road cars look their best with the right height, the right wheels and largely stock bodylines.

'At the moment, my plans for the future include just figuring out how well everything works. It's been great out of the box, but with more kilometres on it I am identifying little changes, like damping, brake bias et cetera. Eventually I'd like to do a few track days, autocross, and sprints in it but it's not destined to be a race car. It's just a fun weekender really to keep me feeling young and stupid every so often. I will go to great lengths though, to keep it looking fresh and tidy.'

As for Richard's photography career, how did that come about? 'I've been doing it seriously for about 10 years. It started as a way to be involved in motorsport when I wasn't able to afford a car, and I'd grown tired of marshalling.

'The highlights are the people I meet doing this. They start as customers, or car owners, but many become friends. I never envisaged the access I've had to so many cars of significance, both culturally and historically. The absolute highlights for me would be getting amongst so many notable old touring cars; the Group A and Super Touring era

is my favourite, and being able to get up close and personal with these machines I'd only ever been able to see on TV as a youngster is something I'll never tire of.

'I think shooting your first cover is probably something that never goes away; in my case, it was a KE70 Corolla we shot on Ruapuna back in 2015! It really felt like an achievement. I could go on and on and on, but I guess in a nutshell, it's a real privilege to be both trackside and in front of some of the coolest cars this country has to offer.'

Rod Harvey

Kiwi drag racer Rod Harvey is one of the fastest door-car (sedan) pilots on the planet. Period. He rose to national prominence with a fast and wild 1000-horsepower Datsun 1200 coupe that became an instant fan favourite, before getting a whole lot more serious with a 2500-horsepower Toyota Celica with which he broke numerous records, including the world's fastest top speed Import, and the first Import in the Southern Hemisphere to go under 6 seconds in the quarter mile. At the time of writing, Rod is in the top six fastest Import drivers in the world. And yet, this master of small-capacity high-tech turbocharged competition cut his teeth in big old American V8-powered muscle cars.

Import, or Sports Compact drag racing is an American term, used to describe high-tech, small-capacity cars, mostly of Japanese origin. It's a simple term to differentiate from the traditional domestic V8-powered drag cars from which the sport originated. Import racing may have been low-key to begin with, but these days it's taken incredibly seriously, and is responsible for some of the fastest cars on the planet. Furthermore, the popularity of Import drag racing has spread to all corners of the globe, including New Zealand.

The Import scene bears many similarities with the early days of hot rodding, in that its numbers are dominated by young men and women who equip their modern daily driver with aftermarket parts that improve the way it looks and performs. Furthermore, each owner works hard to differentiate their car, not just from that in standard form, but from every other modified car. And naturally, when Import drag racing began to gain traction, so these same young owners predominated, much like that of the early days of drag racing.

'I started drag racing when I was around 17 or 18, just out at Meremere,' explains Rod, who hails from Auckland. 'That's predominantly what I did as a teenager. I started with an XA Falcon GT coupe, but I had various Mustangs and Falcons. Back then I had to sell one to pay for the next, but there are some I wish I had kept. I eventually graduated to a 1995 Mustang with a big block Ford in it. At the time it was the fastest street car still fitted with street tyres and mufflers. It would run at 8.80 [on the quarter mile] on street tyres and mufflers.'

Rod then made the switch to competing in the fast-growing Import scene, with a wild Datsun 1200 coupe. It was quite a radical departure from his background to that point. But his reasons for doing so were mostly financial. 'At that time, if you went to the V8 drags at Meremere, there were probably 500 people there. But if you went to the 4 And Rotary Nationals at the same place, you couldn't even get in the gate. There were 15,000 people trying to get in. So if you want to have a sponsor, you had to go where the people are. It seemed to be what everyone was doing, and I looked at it and thought, "you know what, we could probably work some of this stuff out". Not too many people doing the Import stuff back then really knew how to send cars down the track. Between myself and Terry Bowden [Rod's drag-racing partner], we'd been racing a long time, and we knew how to build a tube-chassis car, and we thought we knew how to send a car down the track. So we decided to see if we could make an Import motor work, and figure out how to work an ECU and turbocharging. We were just learning so much, and we just never stopped.

'So we had this little Datsun coupe that Terry built for me in around 2002. We put this thing together between the two of us, we put a turbocharged [Nissan] SR20 in it, and that's when I built my first Motec computer, and figured out how to wire it myself. And then we just evolved from there.

'The fastest I ran in the Datsun was a 7.70 at 183 mph. It made just over 1000-horsepower. It weighed about 900 kilograms. It was a pretty cool little car.'

Rod's Celica launches hard. From this angle the stretched proportions of the carbon-fibre body are evident. After several years of development, this car ran a 5.90 at 239.31 mph at Willowbank Raceway in Australia.

For a drag car, the longer the wheelbase, the more power it's able to handle, and the more stable it is at getting down the track, hence why Top Fuel dragsters, the fastest cars on the planet, have a 300-inch (7620mm) wheelbase. Top Fuel cars no longer race to the traditional quarter mile track length, because top speeds for these 10,000-horsepower monsters were becoming too high, and crashes were sometimes fatal. These days they run a shorter 1000 feet (305 metres) track length, but can still reach 335 mph (539 km/h), with the fastest run to date taking just 3.64 seconds.

After a few hair-raising years in the Datsun, Rod knew that to go faster, he needed something with a longer wheelbase, which, in 2008, is where the Toyota Celica came in. Now things really started to get serious. 'I decided at the time, to move forward from the Datsun. The longer we could make the car, the easier it is to get down the track, and the safer it is. And that was the direction we headed.' Unlike the Datsun, which Terry and Rod built from the ground up, the Celica was purchased as a turn-key race built by Rick Jones Race Cars, in the United Sates. 'Rick Jones, back around 2003, 04, 05, built a bunch of these tube-frame Celicas for the NHRA Sports Compact class. After a few years that class faded away, so a bunch of these cars came up for sale. The cars were based on a Chevy Cavalier Pro Stock car, but in the case of my car, instead of putting a Chevy Cavalier body on it, they put a Celica body on. As a Pro Stock car it would have had a 500 cu in big block Chevy with a single carburettor, and in those days it would have run the quarter in around 6.50 at 200 mph. So those guys know what they're doing, and we knew if we wanted to go faster this was what we had to do.

'The tube-frame chassis was made from chrome-moly. The motor was a [inline-6] turbocharged 2JZ, which typically came out of a Toyota Supra. The car was already fitted with the 2JZ. We went to the States, purchased the car and brought it home, then ended up changing pretty much everything they'd done because we really struggled with it. We ended up putting a Motec in it, and just kept changing things and changing things. I don't think there was one thing we left unchanged from when we first got the car.

'We run a Lenco transmission, which is designed specifically for drag racing. I use a clutch when I leave the start line, but then it locks up and you shift sequential after that. You can keep adding on to a Lenco. You can add another case to give an extra gearset. If you had two cases you'd have a 3-speed, if you had four cases you'd have a 5-speed. I've tried everything, I've tried 3-speeds, 4-speeds, 5-speeds. I've tried every ratio under the sun to try and make these things go fast down the quarter mile. But now I usually use a 5-speed, just to keep the motor in a certain power-band, and keep the ratios all as close together as I can. We use a custom-built differential housing which is based on a Ford 9 inch. We run 9.5-inch gearsets in them. We typically run 16 x 16 inch back rims, with a 17-inch-wide tyre, which is what they run on Doorslammer cars.'

The wheelbase on the Celica was 109 inches (2768mm), which is around the same as an average family road-going saloon, which doesn't sound like much, but for Rod, it made all the difference in the world. 'After the Datsun it felt like a Rolls-Royce. It was so much easier to drive, and so much safer. For the entire period we raced the Celica, it weighed 2420 pounds (1097kg). The class we raced in, in Australia had a minimum weight limit of 2400 pounds, and we always ran the car at 2420.

The Celica body was made of carbon-fibre. And while the new Pro Stock-based Celica was considerably longer in the wheelbase than the Datsun, it was also significantly more powerful. After racing this car for 10 years, and continuously developing it in every area, including the engine, the team was extracting around 2500 horsepower. 'As things stand with my new car [more on that soon], and with the Celica before I parked it up, you need at least 2500 horsepower to do 250 mph when running to that weight.'

Rod began his Import racing career in New Zealand, before making the switch to Australia, where the competition was much tougher. That in turn made him faster, and pretty soon he was knocking on the door of the fastest Import cars on the planet. 'After a while, I started competing against guys from all around the world. In the Import world, you're just number one if you go the fastest. Everyone knows and everyone remembers if you go fast. Nobody seems to remember if you won a championship, or won your race on the weekend. But if you go super-fast, and you're in the top ten [for speed] in the world, then you're the man. That's how it rolls in the Import world. It's a hard case. I won two ANDRA [Australian National Drag Racing Association] Championships in 2011 and 2013 in the Celica, and nobody outside of Australia even knew. But in 2014 I was the fastest in the world; I ran a 6.12 @ 239 mph, and by the time I'd rolled around into the braking area my phone was ringing off the hook, and social media was just out of control. This was from people all around the world, anywhere Import drag racing takes place. When you go fastest in the world everyone knows instantly.'

When Rod first began racing the Celica, his goal was doing quarter mile times in the high 6 second bracket. But he just got faster and faster, as the Import drag scene as a whole became faster. 'Nobody ever thought one of these cars could make a 5 second pass. But when I went 6.12 I started to think, "hey, maybe we *can* make one of these cars do a 5 second pass". Two years prior to that, if you'd even asked me if we could do a 6.1, I'd have said we didn't have a shit-show. So when I did that time, suddenly everyone got on the band-wagon trying to run a 5 second run with an Import car. But it just became a money game. And the guy with the most money was going to win.' Ultimately, that's exactly what happened. Ebrahim Kanoo of Bahrain, owner of EKanoo Racing, funded a Toyota Supra powered by a turbocharged 2JZ, which, driven by Gary White, ran a 5.97 at 240 mph (387 km/h) in 2015,

The new car, rolling on a Pro Modified chassis, and draped in a Toyota Camry body. The carbon-fibre body was originally modelled on a Chevrolet Camaro, but when Rod purchased it and converted it for Import drag racing competition, he changed the nose to resemble that of a Camry.

to become the first Import car in the world to dip into the 5 second bracket. Since then, a handful of Import teams have managed the same feat.

Rod joined this elite group in June 2017, when his Celica ran 5.90 at 239.31 mph at the Winternationals at Willowbank Raceway in Australia. 'I was the first in the Southern Hemisphere and this side of the world to run a 5 second pass, and still the only person on this side of the world to run a 5 second pass in an Import car.' That's a staggering achievement.

The one side-effect of this team getting so fast is that Rod no longer races in New Zealand. He predominantly competes in Australia these days. His kids race Junior Dragsters in New Zealand, but that's about the extent of his domestic involvement now. 'There just isn't the competition here. Even if you look at the Top Doorslammer field, which is for blown-V8s [usually in excess of 500 cu in capacity], there isn't even a Doorslammer that's run a 5 second pass in New Zealand yet. I've done it in an Import car with 190 cu in.'

As impressive as it is for a car with what is essentially a production car engine block and head to be getting down a quarter mile drag strip in these sorts of times, eventually, something had to give, which has forced Rod, and the other top teams also using the 2JZ package, to look for alternatives. And that meant race specific components. 'I run billet engine blocks now, but for a long time I was running the factory Supra block. We stroked it to 3.2 litres, and still to this day we run a factory cast aluminium cylinder head. Back in the day if I was to run the Celica with the cast 2JZ block and factory head, with 50 pounds of [turbo] boost, it'd run 6.70s at 200 mph all day long. You could run a whole season without having to put a spanner to it, really. Just check your bearings regularly, and that was how it went. We've developed

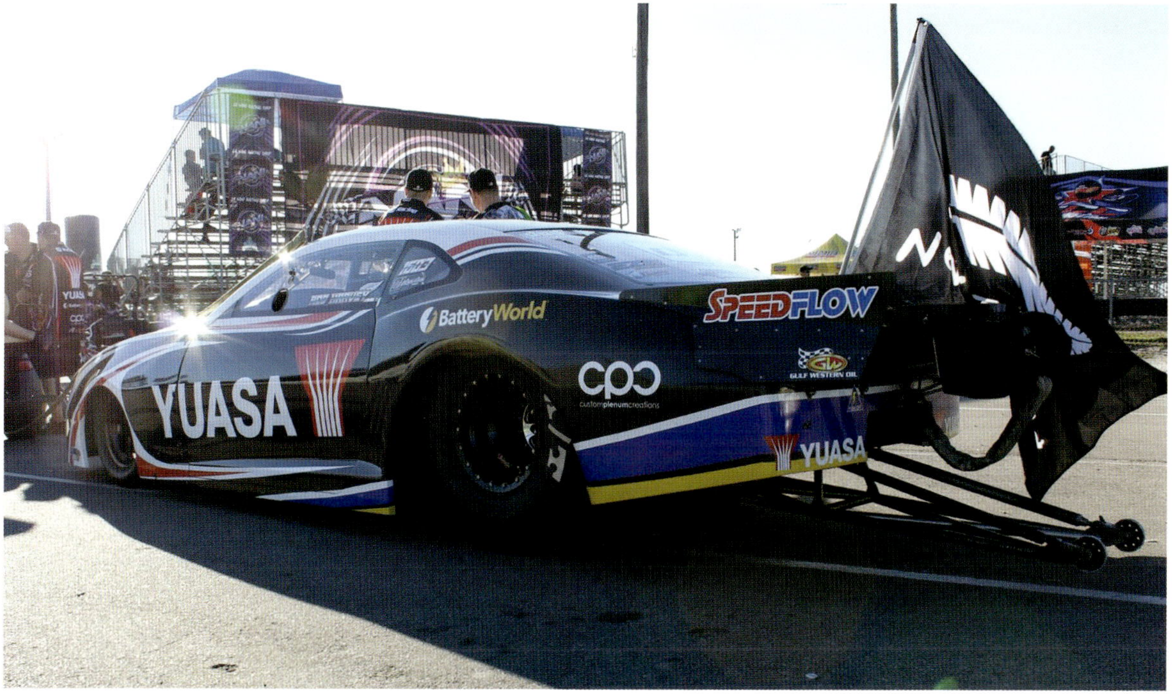

pretty good camshafts, valve-train, and things like that to take the boost. But the faster you go, the more you beat up on the gear. These days we're putting as much as 95 pounds of boost in it. And that's why we need the billet block.'

Obviously, the engines Rod runs are highly developed and highly tuned, and horsepower increases have been gradual. But how do you get 2500 horsepower out of a 3.2 litre single turbocharged Toyota 2JZ motor? The answer sounds simple, but, of course, it isn't. 'You just keep adding more [turbo] boost. Boost is your friend. You just keeping banging in more boost. If you can move the air, you can keep making more power. That's what fascinated me with these things. When I stopped racing V8s and started racing Import cars, I couldn't believe how much horsepower you could get out of such a small motor. But if you look at Formula 1, that's what they do too.'

Incredibly, despite the Celica having run a 5 second pass, Rod could see it was nearing the end of its useful life, and with Import cars getting faster and faster, knew he'd have to step up into something even better. 'Early on when we started running this car, our goal was to run a 6.90. But when we first got it, it wouldn't run properly or do anything right. The guy I got it off in the US had a little bit of success with it. He'd run a 6.50 @ 217 mph once, but I think he just got a little bit lucky that one time. When we first got it home, nothing worked. At first we were running 7.40s, then 7.30s, and our goal was to be the first Import car in New Zealand to run a 6 second pass. And we did that. We took the car to Australia and ran it there for a year or so, and did some 6 second passes there, and went over 200 mph. So we shipped the car back home and were the first to get into the 6s. But then there was nobody here to race, so we just sent the car back to Australia.

'When we bought the Celica it was based on a Pro Stock car, which [the chassis] was designed to run a 6.50 @ 200 mph. I had it for eleven years, and it raced for a couple of seasons prior to that. And I made it go 5.90 at 250 mph. When that car was built it wasn't really designed to go 250 mph.'

In late 2018, Rod unveiled an all-new car, which he races as a Toyota Camry. 'The new car is a Pro Modified car; more like a Doorslammer. It's longer again, and it's designed to run 250 mph. It's got a double-rail chassis [with a pair of stacked rails running front to rear on either side], which the Pro Stock cars don't have. Terry did convert the Celica to a double-rail. The Celica became quite hard to get down the track the faster it got, so Terry converted it to a double-rail [which reduces chassis flex]. 3/10ths of a second is huge in our world.

'The rolling chassis [with 112-inch wheelbase] started life with a Camaro body which we then changed to a Camry by fitting a different nose. That way we could carry across the drivetrain from the Celica, and still call it a Toyota. That's what a lot of the guys in the US have done. In the timeframe I had, going to Orlando [Florida, for the 2018 World Sports Compact Challenge], and knowing that I didn't really want to take the Celica, and that I'd be forever battling the Celica, the easiest way was to buy a Pro Mod, change the nose, and call it a Camry. A lot of other guys have done that as well. The top ten Import cars in the world, which are really the only cars to have gone into the 5 second bracket, are all Toyotas, all 2JZs.

'When we showed up in Orlando with the new car, I hadn't even made a full pass. I managed to get the car going. I had one meeting in Australia to test the car before we put it into the container to go to Orlando. I made it to half track and it was struggling with some stuff, but it was showing really good signs of going fast. The chassis was working really well, but it was raining that weekend, and the following week I had to load it into the container. So when I got to Orlando, it was the first time I'd done a full pass in the car. You could imagine what was going through my mind. But we had a great time, and we ran back-to-back 5.85s @ 254 mph [faster than he'd ever been in the Celica].'

Part of the appeal for Rod in running the new Pro Mod chassis was that, not only was it designed for faster runs than that of the Celica, it was also lighter. Although Australia has a minimum weight limit, the Celica was as light as it was ever going to get. But other parts of the world where Import drag racing takes place run a lighter weight limit. The US run to 2200 pounds (997kg), and virtually all of the cars that have got into the 5 second bracket are running to that 2200-pound limit, including the EKanoo Racing Supra. Rod's new Camry provides the option of being closer to the 2200-pound limit for competing in the US, but weight can be added to race in Australia. In Orlando, the Camry weighed 2300 pounds.

Amazingly, even though he went faster than he'd ever been with the Celica, Rod still wasn't the fastest in the world. In fact, that 5.85 put him at number 6. The fastest Import car at the time of writing had done a 5.66. 'The guy who ran the 5.66,' recalls Rod, 'he decided to add another 400 horsepower using nitrous to the turbocharging. He went 5.66, then the next day he had a nitrous explosion and the car burnt to the ground.' And thus begins the ongoing battle of trying to go faster again, while others are doing the same. It's a constantly moving set of goal posts, and even as this is written, it's probably already out of date, such is the ongoing evolution of drag racing. Rod is sure this car has a 5.7 in it. By the time this book hits the shops, it's very likely he will have achieved that.

'It's an out-of-control hobby for me. I have some really supportive

sponsors, and I couldn't do it without them. We're doing some nice stuff here with CNC machining, but I go to the States and I look at some of these cars, and think "wow". But I look at the gear that they've got, the gear they're using, the tracks they've got, and I think the tracks they're racing on has a lot to do with how fast they are. And I look at it and think, "you know what, I could still be number 1 again". But at what cost?'

Drag racing is a unique form of motorsport in which a person constantly strives for the perfect conditions; atmospheric, track, and car setup, in the hope that everything aligns and they can burst down a short, quarter mile drag strip in a few seconds, in the quest for a time that's a millisecond or two faster than they've been previously. And there comes a point where, they're going so fast, and the gains have become so minute, that most of their attempts are slower than their best. No doubt, for every great achievement, must come many more frustrations and near misses. But the thing that keeps them going is the belief that, sooner or later, they're going to go faster than they've ever been previously, and perhaps, faster than *anyone* has been previously. It's an addiction. But such is the life of a drag racer.

Sarah Campbell's *Cressida*

The Toyota Corona Mark II was introduced by Toyota in Japan to be positioned between the high-end Crown, and the slightly smaller Corona.

Although purely a domestic model to begin with, Toyota began selling the Mark II in the United States in the early 1970s, taking advantage of the global oil crisis, as US buyers turned away from buying big American offerings.

When Toyota launched the third-generation Corona Mark II in 1976, it rebranded all its export models as the Cressida. Here in New Zealand, assembly of the Cressida began with the third-gen model, where it replaced the larger Crown as Toyota's most luxurious offering.

The fourth-generation Cressida arrived in 1980, sporting the latest European styling, with sharp lines and large glassware with thin pillars.

Historically, luxury models have a tendency to become highly undesirable once they're a few years old. Their larger proportions tend to age them, and the used-car market doesn't favour them. As such, their survival rates are generally quite low. It's rare, therefore, to find such a beautiful and well-preserved Cressida as this fourth-generation example, owned by Sarah Campbell. Despite having been around the clock several times, it is a quite stunning car.

'This is my wife Sarah's RX60 Cressida. She is very proud to own it. We've owned it for around three and a half years and bought it off the original owner who traded his HQ Holden to buy this Cressida brand new in 1982.

'In that time the original owner managed to clock up over 480,000km and currently we're trying our hardest to hit the magic half million. Only about 7500 to go! It still has the original engine, with no major repairs ever being completed. Has all the original sales documents and we were told that only two people had ever sat in the back seat, which was for a wedding. It even has a bare metal patch right above the driver's door handle where the original owner used to rest his thumb to close the door after locking it.

'We managed to purchase it off him as he traded it in at my work after his wife told him he needed to upgrade. He now drives a late model Mercedes Benz C Class.

'Since owning it, all we have done is kept it maintained and professionally groomed, added some Yokohama period wheels and lowered the suspension a little bit.

'My wife is unaware her car is being featured in this book. She'll be over the moon.'

Jared Campbell

Scott MacDonald's *Corona*

When Toyota launched the fifth-generation Corona in 1973, its range of engine options spanned 1600cc–2000cc, with its most powerful boasting just 88 horsepower (approximately 65kW).

Today, the classic fifth-gen Corona is a popular choice among enthusiasts, who gravitate towards its beautiful styling. But while some owners opt for pure stock, as they rolled off the assembly line, Scott MacDonald's 1976 example is right at the other end of the spectrum.

'It all started for me when a couple of friends started to get into the older Japanese cars, and after a bit of searching on TradeMe, I came across this gold 1976 Corona. I had never actually seen one before and from the first time seeing the photos, I had decided that was it.

'I've always been a Toyota fan and there were plenty of Corollas out there, and I've never been a major fan of the rotary scene, so the gold Corona with a tired old 18R 2-litre carburettor engine and 3-speed auto was the perfect choice.

'I ran with the factory setup for a while but after taking a road trip to Auckland and having it use 4 litres of oil on the way I knew it wasn't going to stay like that for long as I have a bad habit of having to tinker with things to make them better.

'So what better way to give the old Corona a new burst of life than to repower it with the more modern 3SGE BEAMS 2-litre and 6-speed manual gearbox from an Altezza. This was a pretty straightforward swap compared to my last project, a 1995 Hiace van with a 2.5 twin turbo 1JZGTE. That thing was a weapon.

'And of course I couldn't just leave the 3SGE the standard normally aspirated setup. So a Turbonetics turbo soon found itself at home, and sitting alongside it among other upgrades such as a big set of Willwood 4-pot brakes, link G4 ECU, and Recaro front and rear seats from a Daihatsu Charade of all things. They fit almost perfectly. The Corona sits on a set of 17 x 8 inch rims in front and 17 x 11 inch rims in the back.

'It all involved countless hours and late nights in the shed and several dollars, but it all went together pretty good. And when it was dyno tuned on 10psi boost it put out a nice 210kW at the wheels. But that number will definitely be going up in future. Being a machinist engineer by trade certainly helped with making plenty of custom parts as there is pretty much nothing off the shelf for this car.

'After driving it for roughly a year it was pretty apparent that the old diff with leaf-spring suspension was not loving the extra power and gave quite bad axle tramp when you put your foot down. It sounded like it was going to jump through the back seat at you. So that all needed to be dealt with but, after crunching numbers and the engineering logistics, a 4-link Hilux diff put me off a bit. Luckily enough, my father, who used to muck around with hot rods, suggested putting a Jaguar independent rear suspension setup into it which seemed very odd to me, but on further investigation would work out to be a good option. It was very easy to fit in with minimal modifications to the Corona, and was solid enough to take the new power going through it. The other advantage was inboard 2-pot disc brakes that made for an unusual view through the back rims, not seeing any brakes at all.

'Then came a full set of QA1 adjustable coil-overs front and rear to give it its final ride height of 75mm off the ground without the rock-hard feel of sitting on bump stops. It made the car handle so much better and with a bit of camber on the rear wheels gives it a stance I'm happy with.

'As it's still a work in progress, like many cars, eventually the gold will be swapped out for Voodoo Blue and the engine will be getting some nice forged goodies inside so it can handle some more boost and get into the 320kW (approximately 430 horsepower) zone.'

Shane Harwood's Superstock

This incredible machine is a Superstock speedway car. Not only are Superstocks the most popular form of speedway racing in New Zealand, they're one of the most popular of all motor racing categories here. Furthermore, Superstocks are one of the longest continuously running racing classes in New Zealand.

Superstocks have their roots in stockcar racing, which began back in 1954. Speedway racing itself began in New Zealand in 1929, for Solo motorcycles. Midget racing was introduced officially in 1937.

As the name implies stockcar racing in New Zealand was for virtually stock-standard cars. The same was true for the sport in the United States. However, each country has seen its sport evolve in completely different directions.

New Zealand stockcar racing started at Western Springs Speedway, which opened in 1929. Such was the grassroots nature of the category, stockcar races were held after the regular speedway season ended. Unlike American stockcar racing, the New Zealand variant encouraged the use of front bumpers to manoeuvre opponents out of the way. This in turn required the cars be fitted with stronger bumpers, which slowly evolved into them eventually becoming a purpose-built steel frame on which a cut-down body was mounted. In the early days, the body was a road car shell shortened, narrowed, and with its floor and firewall removed.

There are 23 Speedway New Zealand-recognised tracks in the country, from Auckland to Invercargill. Each is approximately 400 metres in length, when measured along the pole line, and although they're all of an oval shape, each is unique in its own way.

Speedway racing ensures that each track has its own contracted drivers, who race at that track throughout the season. There are also multiple classes, each of which has multiple championships to contest each season. The most coveted of these is the New Zealand Championship. But there are also North and South Championships, and a New Zealand Grand Prix. Championships are contested in just a single event, held each year at a different track. Most championships are over-subscribed, requiring a qualifying night, in which the field is broken down into groups, and each group has three races. Starting positions are drawn at random, then rotated for each race so each driver starts a race from the front, middle, and rear of the grid, to provide a level playing field.

Most classes only allow the best 20 cars through to the Finals night, but the full contact classes allow 26. Most championship Finals nights then have three races, with the driver scoring the most points being awarded champion, although sometimes a single Finals race is contested instead.

The first New Zealand Stockcar Championship was held in 1966, at Palmerston North Speedway, and won by Graham Stretch. This category has been incredibly hotly contested since its inception, and race fans are extremely loyal to their local track and favourite driver. This is very much a generational sport based on tradition.

During the 1990s, as the Stockcar class became ever more competitive, so teams were spending more and more money on building the ultimate racing machine. This saw numbers begin to decline. However, a new class, using older Stockcars with less powerful engines, was created, which then had the added advantage of becoming a feeder class. Initially, the new class was called Standard Stocks, but eventually, was renamed Stockcars, while the Stockcar class was renamed Superstocks.

Modern-day Superstocks are impressive machines, that are completely unique to New Zealand. Top cars can cost well in excess of $100,000. They're built from a steel chassis, most of which use a space-frame design. The chassis is designed to be both fast and agile, but

sustain heavy contact. This is a full-contact class, where eliminating the opposition through virtually any means is required.

Superstocks are race cars. Just like any form of racing, they compete over a fixed number of laps, and the first driver to complete that distance is the winner. But the way in which they beat their opponents is what makes this category so unique. Although built to hit and be hit, these are very fast race cars.

The Superstock category has a maximum engine size of 248 cu in. The last two decades have seen the regular use of Japanese 32-valve quad-cam V8s from Nissan and Toyota used in Superstocks. The Toyota engine is the 1UZ-FE model, measuring 242.1 cu in in standard form, and which was originally produced for powering luxury Toyota models, including the Lexus LS 400 and Toyota Celsior. Second-hand, these engines are relatively cheap, and imported in bulk into New Zealand.

The 1UZ-FE is factory fitted with fuel-injection, but this has to be removed, as Superstock rules only allow for a single 4-barrel carburettor. In addition, they're completely rebuilt for racing, and there are several specialist builders around New Zealand who prepare them specifically for speedway teams, including Bryan Hartley, father of Brendon Hartley. It's not just Nissan and Toyota V8s being used, others favour Ford and Chevrolet engines. The very top Superstock race motors can produce around 500 horsepower, pretty impressive given their relatively small size.

Superstock regulations require a minimum racing weight of 1400 kilograms. However, this is one of the few motorsport categories that also has a maximum weight, which is 1500 kilograms. The reason for this is purely for safety. Being a full-contact sport, if two cars hit each other and one is far heavier than the other, the results can be disastrous.

Superstocks use 15-inch-diameter wheels on all four corners. Even though they only turn left, all four wheels are the same size. However, cars are heavily weight-jacked to bring as much weight across to the left as possible which helps cornering. This includes shocks and springs which are often higher on the right side as is tyre pressure. These cars use a Hoosier treaded racing tyre. The left rear is set to a very low pressure, often as little as 8–10psi. The engine and driveline is also mounted in the left of the car. This is not compulsory, rather by choice.

Superstocks are fitted with a fibreglass body, bonnet, and aluminium wing. The wing does provide downforce, and the regulations limit a maximum size. Various companies produce a range of different bodies, but the most popular is the classic Ford Y.

Superstock chassis are incredibly strong, and are made from either mild steel or high tensile steel. They have virtually no give when making contact with another car, or the concrete wall that lines the track. So while high tensile steel allows for a slightly thinner gauge, which in turn makes the chassis lighter, allowing teams to add ballast in the best areas for performance, the driver's body sustains a greater impact as the high tensile doesn't have as much give. But even cars built from mild steel have very little give. The front and rear bumpers are hollow, so a length of timber can be fitted inside them. This provides a small amount of cushion. But not much. This is an extremely violent sport, and injuries are common.

The Superstock featured here is owned and raced by Shane Harwood, of Nelson. Shane is one of the most successful Superstock drivers in recent years. He won the 2016 New Zealand Superstock Championship, finished second in the 2010 New Zealand Grand Prix, has won the South Island Superstock Championship three times, finished second once, and third twice. Following his New Zealand Championship, he has contested only partial seasons, but won the 2018 South Island Championship in just his second event of the season. He was just 28 years old when he won the New Zealand Championship.

Shane's Superstock was designed by fellow Nelson racer Brendan Higgins, who himself has also won numerous titles. The chassis was built by Brendan and Shane. Shane started racing speedway at the age of 16, competing in a Honda CRX in the Production Saloon class. The Honda had been raced by Shane's older brother Carey, who stepped up to a Superstock. After just one season Shane himself made the switch to a Stockcar, powered by a Ford 6-cylinder built by Bryan Hartley. After just a handful of events, he was suddenly thrust into Superstock Teams Racing, the most brutal of all speedway racing. Teams Racing pairs off two teams from different tracks, each team containing four cars. The race is contested over a selected number of laps, and each team has designated fast cars who try to win the race, and designated hitters, who try to stop the fast cars. The biggest annual speedway event in New Zealand is the Superstock Teams Championship, held at Palmerston North. This event draws teams from all over the country, and is hugely spectacular. But as well as the Palmerston North event, Superstock Teams races are contested on a smaller level at various tracks throughout the season.

Each team has a name; Palmerston North Panthers, Rotorua Rebels, Bay Park Busters, Auckland All Stars, etc. Nelson's team is the Nelson Tigers. Tracks that have a strong field of cars often have more than one team. Shane became a member of the Nelson Tigers B team in just his second season driving a Stockcar.

When Carey Harwood upgraded to a new Peter Rees-built Superstock powered by a quad-cam Nissan V8, Shane took over Carey's old Superstock, which was a 6-cylinder car with a Brendan Higgins chassis. Shane raced two seasons with the car, before Brendan Higgins suggested it'd be better with a V8. To that end, Shane had Graham Morrison, in Blenheim, build him a quad-cam Toyota V8. Shane raced the Higgins chassis with Toyota V8 until 2012, when the decision was made to build a new chassis. This is the chassis built by Brendan and Shane. By now, Shane was a member of the Nelson Tigers, and had developed into one of the fastest Superstock drivers in the country. It was with this car he won the 2016 New Zealand Superstock Championship, held that year at Stratford Speedway.

Richmond
1NZ
Super Liquor
nac

itm
1NZ
NELSON
MOTUEKA
TAKAKA
"We'll see you right"
HIGGINS
WAIMEA
1NZ
TOTAL
bluebridge

itm
18N
NELSON
MOTUEKA
TAKAKA
"We'll see you right"
WAIMEA
18N
TOTAL

HIGGINS
RACE CARS
itm
1NZ
NELSON
MOTUEKA
TAKAKA
"We'll see you right"
WAIMEA
1NZ
TOTAL

He defended his title in 2017, but the championship that year was held in Palmerston North, which boasts the strongest field of Superstock machines in the country. Although Shane was travelling up to Palmerston North most weekends to race, when it came to the championship, the local drivers worked together and, not surprisingly, the top three finishers were all drivers contracted to Palmerston North.

Shane's Superstock runs a Toyota 1UZ-FE quad-cam 32-valve V8, which has been taken up to the class maximum 248 cu in, and produces 470 horsepower. It retains the standard Toyota crankshaft, which has been lightly machined, and uses Manly pistons and conrods. The intake manifold is custom made by Midwest Motors, as is the dry-sump unit. It runs a Brasswell carburettor. Shane has used this same engine in various configurations for the last 10 years, and attributes much of his success to it.

The Toyota V8 is backed by a custom-built 4-speed gearbox, built by TTI Racing in Nelson, and a Winters quick-change differential. The diff design used by most teams uses the Ford 9 inch crown wheel and pinion with straight cut gears. Incidentally, Shane usually only uses two gears. Superstock races feature standing-starts. Shane starts in second gear, then changes up to third gear once he gets up to speed, and remains in third. He sometimes uses fourth, depending on track conditions. Brakes are Wilwood.

Superstock speedway racing is one of the most brutal, exciting, and dynamic forms of motorsport, anywhere. It's fast, loud and highly unpredictable. At New Zealand championship events, it's common to have around one hundred cars all vying to make the top 20 for the Finals. The three Finals heats are electrifying, and hugely atmospheric. To win the championship requires a combination of speed, skill, reliability, and a little luck. After the second heat, when the points are tallied, those out of the running usually go into combat to help their club mates. That's when the action really happens, and cars are rolled, hung on the wall, or just eliminated by any means. That this is one of New Zealand's most popular forms of motor racing is no surprise. It's a little like rugby, but with 500-horsepower race cars.

Shaun Gregory's Starlet

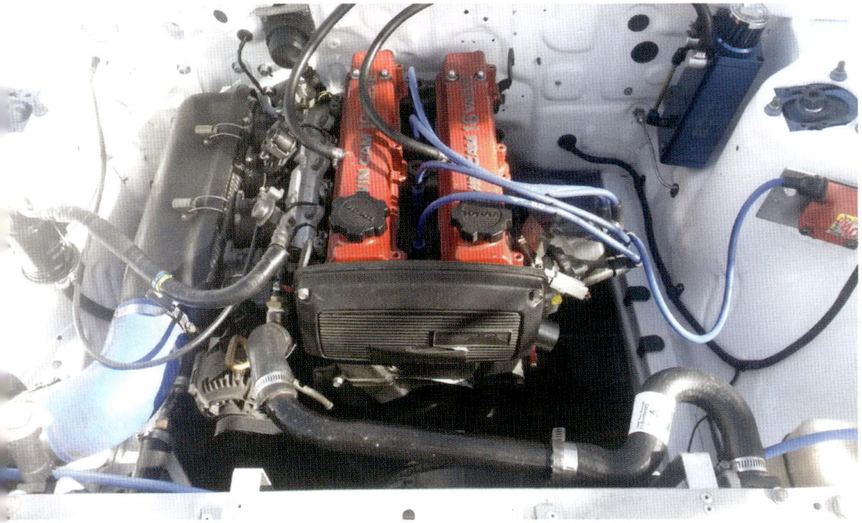

Touring car racing has always been massively popular in Japan, and no more so than in the 1970s, when most manufacturers had some sort of involvement.

The Japanese Touring Car Championship in the 1970s featured several classes, all based on engine size. While the big-bore classes boasted the Nissan Skylines, Mazda RX2/3s, and Toyota Celicas, one of the most popular categories was the 1300cc class, in which hordes of screaming Datsun 1200s and Toyota Publicas would form a long slip-streaming train, with intensely close racing.

And many of the manufacturers had technical input in more than one of them. The grids were massive, and each class would have its own race, which played out dramatically in front of huge crowds. The majority of major races in Japan took place at Mt Fuji Speedway, which featured a colossal 180-degree banked first turn, which the cars hit flat-out off the end of the incredible 1.5-kilometre-long Start/Finish straight.

Toyota was hugely supportive of the Publica in Japanese Touring Car racing, to the point it homologated a wild aero-package, consisting of front and rear spoilers, massive bolt-on wheel flares, and a spree of upgraded mechanical components.

When Toyota launched the second-generation version of the Publica in 1978, it rebranded it the Starlet. Like its older sibling, the Starlet received a series of impressive competition upgrades, and enjoyed significant racing success.

The Starlet was a popular competition car in New Zealand, but more so in rallying than circuit racing. There was, however, a Starlet that competed in the popular 2-litre ShellSport series in the 1980s. And it carried the Toyota factory aero-package. That same aero-package has since graced a variety of Starlet circuit racers, including that of Shaun Gregory, who uses his as a track-day car.

'This is my KP Starlet. It has been to the Toyota Festival at Hampton Downs.

'It runs a 16-valve 4A-GE red top engine. I'm currently fitting a J160 6-speed Altezza gearbox.

'It has had a recent respray in Toyota 040 white.'

Stephen Samuelu's Corolla

The fifth-generation E80 Corolla, introduced in 1983, was the first to carry a front-wheel-drive layout, with the exception of the AE85 and AE86 Sprinter/Levin.

Despite the changes, the Corolla continued to be a popular choice for racers, at a time when the international FIA Group A touring car formula was just gathering strength.

Today, fifth-generation Corollas continue to be a popular choice among Toyota racers, and remain just about the best bang-per-buck on the track. Stephen Samuelu enjoys spirited exercise in his 1985 example.

'This is my 1985 Toyota Corolla AE82. I bought it a few years ago to do local street sprints, autocrosses and hillclimbs. I've since been competing in the Mitre 10 summer series circuit racing at Manfeild and Taupo which I really enjoy. It's just like going go-karting.

'The car started life with the 16-valve 4A-GE engine but has since been fitted with a forged 7AGTE 20-valve engine [7AFE block and Blacktop head) making it a 1.8 litre turbo-charged AE82.

'It's all certified and road legal so it's still able to be driven to events and even to work if needed!

'I love the interchangeable nature of the Toyotas. Most suspension is from the AE111 model, makes it affordable and handle well.'

Team Toyota Rally Corolla

By the early 1970s, Toyota had only been in the New Zealand market for a handful of years. Indeed, Japanese cars as a whole were still something of an unknown, and a rarity. Kiwi buyers were purchasing British, European, and Australian cars. When it comes to car ownership, most people tend to gravitate towards what they know, and what they trust.

To make its mark in the New Zealand marketplace, Toyota New Zealand needed to prove the reliability and performance of its product. And with a ready-made crowd of enthusiasts, the perfect place to do this was motorsport competition.

To that end, Consolidated Motor Distributors (CMD), who were the New Zealand Toyota importers, chose rallying as a platform for which to demonstrate the quality of their products. Rallying was still a very young sport in New Zealand, with the 1969 Shell Silver Fern Rally being the first major event of its type. Mike Bitossi, CMD's production manager, represented Toyota by entering a standard Corolla KE11, and despite boasting just 1077cc, finished an impressive fifth overall.

CMD entered two cars in 1970, a Corolla KE11 for Peter de Lore, while Bitossi drove a rare Corona 1900SL hardtop. De Lore again demonstrated the quality of the Corolla by finishing fifth, while Bitossi placed 21st, after being delayed by an excursion. For 1971, a pair of Corolla KE20s were entered for Colin Taylor and Graham Carter to drive. These cars were lightly modified, but still only had 1166cc engines. Taylor finished seventh, and while Carter was third at the halfway point, dropped back to 36th when he and team-mate Grahame Brownlie had an engine failure.

For 1972, Toyota stepped up to a three-car team, running KE20 Corollas once more. This time they elected to contest both the Silver Fern Rally and the Heatway Rally. Again, the engines were mildly tuned, and Colin Taylor's car had its engine capacity increased, from 1166cc to 1293cc. Ian Pinkerton drove the second car with standard capacity. The third car, driven by Graham Carter, ran a standard capacity engine in the Heatway, but this was increased to 1293cc for the Silver Fern.

In the Heatway Rally, Taylor finished an impressive second outright, to Andrew Cowan's Mini Clubman GT. Carter was thirteenth after several off-track excursions, while Pinkerton rolled his car into a ball.

In the Silver Fern, Carter finished fourth outright, and first in the 1300cc class, while the other two cars suffered gearbox trouble early on, and spent the rest of the rally working their way back up the order, eventually finishing eleventh and twelfth.

Interestingly, CMD were enthusiastic about contesting rally events, but really had no involvement in circuit racing. Perhaps this was because rallying didn't rely on pure raw speed. Consistency and reliability were also factors. And this was where the little Toyotas really excelled.

However, for 1973, the Toyota effort got a lot more serious. When Ford released their new Escort in 1968, they quickly followed up with a performance variant of the little compact, utilizing the Lotus twin-cam engine from the MkI and MkII Lotus Cortina. The Escort twin-cam became an instant success in both circuit racing and rallying. The Escort twin-cam was then replaced by the Escort RS1600, which was fitted with a variant of the Kent-based 1601cc engine using a 16-valve Cosworth-developed cylinder head, and named the BDA (Belt Drive A-series). The RS1600 picked up where the twin-cam left off, and by the early 1970s, was a dominant force in rally events around the world.

Meanwhile, CMD decided they were no longer content with class victories, and began a programme to tackle the 1973 Heatway, with a view to taking on the mighty Escort RS1600s, head to head. To

that end, they imported three brand new Toyota TE27 Levin coupes. These were models not available in New Zealand. Indeed, they looked much like a KE25 SL coupe, which was sold in the domestic market, but beneath the skin were quite different. At the time, they weren't marketed outside Japan.

The TE27 Levins came factory fitted with 1588cc twin-cam engines which, in standard form at least, put them on a par with the Escort RS1600. In standard form, the Levin had 115 horsepower, compared to 109 of the old Escort twin-cam, and 120 of the RS1600. Team Toyota co-driver Graeme Brownlie tested one of the three Levins in standard road-going form for *Motorman* magazine, before it was converted for rally duties. He took the car to Hamlin & Charles, a tuning outfit in Lower Hutt, where they put it on a dyno. They found it peaked at 84 horsepower at the wheels, at 6000rpm. An Escort RS1600 also peaked at 84 horsepower at the wheels, but achieved this at 6500rpm. However, the Levin produced 60 horsepower at 4000rpm, at which point the Escort could only achieve 45.

The Levin's twin-cam came standard with two side-draught twin-choke 40PHH Mikuni-Solex carburettors, and was backed by a 5-speed gearbox, offering one more gear than the New Zealand Corollas. The Levins also had a limited slip differential. Other upgrades included quicker steering ratio at 2.75 turns lock-to-lock instead of 3.3 turns, leather rimmed steering wheel, harder shock absorbers and spring rates, and a larger front roll bar, to improve cornering. Whereas the New Zealand Corolla KE25 coupes had 7.9-inch front-disc brake calipers, the Levins had 9-inch calipers, with power assistance.

The three Levins rolled off the assembly line in Japan painted Burnt Orange, with black interior. Outwardly, they differed from the New Zealand KE25 coupes in that they were fitted with 13-inch-diameter wheels, as opposed to 12-inch wheels. In addition, they featured bolt-on fibreglass wheel flares. And as with Japanese domestic cars of the era, they had the side mirrors mounted on the front mudguards.

The three Levins had to be road registered, and CMD arranged sequential registration numbers for them: GG 41, GG 42, and GG 43.

In preparation for the Heatway Rally, which was contested in the depths of winter, the three Levins were stripped and rebuilt. The engines were modified and included competition camshafts, different carburettors, and a replacement exhaust system. The standard 4.375:1 diff ratio was replaced by a 4.8:1, while the suspension was further

upgraded. The cars were fitted with a rollcage, driving lights, and wider lightweight wheels wrapped in 225-13 tyres. Guards were also attached to protect various mechanical components as well as the radiator.

Once the fabrication work was completed, the three cars were painted a three-tone brown, white and black colour scheme. The driver and co-driver combinations were Colin Taylor/'K' Foulkes (GG 41), Graham Carter/Graeme Brownlie (GG 42), and Ian Pinkerton/David McKinney (GG 43).

Rally events in the 1970s were arduous and gruelling contests, and the Heatway was no exception. In 1973, it measured a total distance of 5074 kilometres, 2812 of which were Special Stages. The rally took seven days to complete. For the first time, it included both the North and South Islands, and competitors were confronted with various challenges, including snow and ice, night stages and river crossings, as well as the usual dirt and gravel surfaces. Just to make the distance was an achievement in itself.

The rally started in Christchurch, then continued to Akaroa, Dunedin, over the Crown Range, through Haast, Greymouth,

TINKLE TINKLE LITTLE LEVIN

CONSOLIDATED Motor Distributors — the national Toyota importers — have always been keen on rallying their products as a means of demonstrating their speed and durability. This goes back to the first true rally in this country, the 1969 Shell Silver Fern, when Mike Bitossi finished fifth overall in a standard Corolla KE11 (1077cc).

Peter de Lore followed this up in the 1970 Shell Silver Fern by taking another fifth overall (again in a standard Corolla KE11), while Mike Bitossi fielded a rare Corona 1900SL hardtop but could only manage 21st place after an off-road excursion.

In the 1971 Heatway, two Corolla KE20s were entered, one driven by Colin Taylor (finishing 7th overall) and the other by Graham Carter and myself (36th, although we had been third overall at halfway ... poked a rod through the block) ... were fairly midly modifi... 1166cc but ... horsepower ...

Story by GRAHAME BROWNLIE
Pictures by
MARSHALL PHOTOGRAPHICS

In the Heatway Taylor took an excellent second place to Andrew Cowan's Mini Clubman GT, Carter/Brownlie finished 13th after two trips into the boondocks, and Pinkerton retired after a massive roll. Then in the Golden Shell Carter/Brownlie finished fourth overall and first in the 1300cc class while both our team mates had gearbox troubles early on and had to claw their back through the field to finish ... overall.

In each rally ... Landcru...

after that town in NZ with the racing circuit; the word apparently means "Lightning."

The photographs show that the Levin looks almost identical to the SL apart from bigger wheels and the screw-on flared guards. But the difference — and it *is* a difference — is under the bon... lurks a most-impressive-... cam engine.

Yes, the ...

Because the car is based on the Corolla coupe SL which was tested by *Motorman* in February 1971 — and which is familiar to most readers anyway — I'll concentrate on the differences. Externally, as I've said, the differences are few — flared guards, bigger wheels (5J x 13) and three "Levin" badges (one on the bootlid and one on each side of the car, behind the front wheels).

The car I tested was fitted with Bridgestone 175 x 70HR13 tyres, although the other two were on Yokohamas. The bigger wheels certainly fill the wheel arches much better than do the 12in ones on all other Corollas, but for the rally the wheels will be replaced by still wider alloy ones and fitted with 225 x 13 Goodyear Ultragrip "knobblies".

At a glance, the interior is similar to an SL, but in fact there are quite a number of detail differences. For example, there is an extra row of gauges in the centre of the dash — oil pressure, oil temp, water temp and ammeter — and all instruments are metricated.

The oil pressure gauge reads in kilograms/square centimetre and on Carter's car sat on 3½, which doesn't sound right but is actually 50psi; the water temperature is in degrees Celsius; the speedo and odo are in kilometres. Thank goodness rpm don't have to be converted, it's nice to have something familiar to hang on to. The tacho has an orange segment from 6500 to 7000, where the red segment starts. The speedo would really impress small boys, because it reads to 210 — but that's kph, which converts to about 130mph.

Other interior changes from the SL include a footrest to the left of the clutch pedal; the gearlever knob (leather covered and having a diagram on it showing *five* gears); the steering wheel (leather-rimmed and having two small horn buttons on each ...); the seats are similar but have large ventilating holes (which really work!); the rear window is electrically heated; and the centre console is a bit different and includes provision for a clock, radio and cassette player.

Moving on to the mechanicals of the car, the suspension setup is basically the same as the SL, too, being McPherson strut at the front and semi-elliptic leaf springs with telescopic shocks at the rear. However the spring rates, front roll bar, front strut bars and shock absorbers have all been uprated. A limited slip diff is standard wear, but for the rally the 4.375:1 units will be replaced with 4.8:1 jobs for better acceleration. The brakes are beefed up also, the normal 7.9in front discs being supplanted by 9in units with power assistance. The booster is a good feature, because without it hard braking in a Corolla requires a hefty push.

The steering has received some attention as well, being 2.75 turns lock-to-lock instead ... required by the standard ...

So you can see there are a lot of similarities, extending even to the unladen weight which is 16.7cwt for the Levin and 16.4cwt for the Escort twin cam. Overseas reports give the Escort a standing quarter time of 16.9sec and an identical time for the Levin.

But what is it like to drive? Basically very much like any other Corolla as it is still a basic understeerer, but the extra power is fantastic. And guys who think they're dragging off a Corolla SL from the lights sure get a surprise! Even more impressive than the power, though, is the torque available from the willing motor. It is really quite incredible, because even in the overdrive fifth gear (pulling 0.86:1) it will pull away smoothly from as low as 2000 revs — even going uphill!

The balance of the car was — to my mind — not quite as good as an SL because the twin cam engine and 5-speed box add 150lb extra weight to the front of the car, and even though the slippy diff is 25lb heavier than a normal one it still leaves the front end too ... to an SL. This, combined ... the steering

Murchison, Blenheim, and back down to Christchurch. The teams then boarded the *Rangatira* ferry from Christchurch to Wellington. From Wellington, they continued to the Wairarapa via Paekakariki and Moonshine, then to Palmerston North and Hastings. They then went north to Gisborne, Opotiki, Rotorua, Hamilton, and on to the finish in Auckland. Much of this was on unpaved roads.

The Heatway Rally attracted not only international entries, but also manufacturer involvement. This was essentially a showcase for car manufacturers. It boasted an impressive 120-car entry list, including hot favourite Hannu Mikkola, of Finland, who'd go on to win the World Rally Championship in 1983. Mikkola was entered in an Escort RS1600. Australian driver Colin Bond was in a Holden Dealer Team Torana XU1. Bond won the Australian Rally Championship three times. Shekhar Mehta was entered in a factory-supported Datsun 180B.

Escort RS1600 drivers were plentiful, and included Mike Marshall, Jim Richards, Blair Robson, amongst several others. Rod Millen was entered in a Mazda RX3, while his brother Steve had a Torana XU1.

A television documentary, directed by Tony Williams, followed the 1973 Heatway Rally, in which the three Team Toyota Corollas featured prominently. However, their rally would come unravelled virtually from the outset. On the first stage, the Carter/Brownlie Levin struck its first clutch failure. By the time they reached Timaru at 9pm that night, they'd had five clutch changes. And it was an ongoing series of clutch failures that would beset all three Levins. Taylor/Foulkes were slowed by similar failures in the first two days, and had lost so much time they were already out of the running. Only the Pinkerton/McKinney Levin escaped without a miss, and was running strongly early on. However, Pinkerton ploughed too quickly into a ford crossing along the Hakataramea stage, and bent an axle. Co-driver David McKinney later became extremely ill, and had to wrap himself in a sleeping bag for a couple of stages before eventually requiring hospital treatment. Eventually, the car also struck clutch trouble.

After the first 1000 kilometres, Taylor had gone from fifth to 105th, while Carter was four places further back. Taylor's Levin required an engine change, and was fitted with a standard twin-cam for the rest of the rally. But despite all the delays early on, he was proving the Levin had potential. On a stage near Haast, he placed fourth fastest outright, despite only having a standard engine.

In the North Island, the three Levins suffered far fewer problems. Indeed, they were among the top three fastest cars on several stages. But they'd lost so much time early on, they never clawed it back. Small dramas, mostly relating to minor incidents, provided further delays. When they eventually reached the finish in Auckland on Saturday, 14 July, they were 74th (Taylor/Foulkes), 78th (Carter/Brownlie), and 91st (Pinkerton/McKinney).

However, they'd shown that they could have been contenders. Perhaps not against Mikkola, who was the class of the field, and finished a full 40 minutes ahead of second-placed Mike Marshall, but certainly the Toyotas should have been competing for top five placings.

Following the 1973 Heatway, all three Levins were sold, after which they continued their rallying careers. The subject of this story is the second car, GG 42, which is the same car featured in *Motorman* magazine in standard road-going guise prior to being rebuilt for rallying. GG 42 was sold to Ken Cleghorn, who rallied it for several years, before it was sold to Ken Adamson. It was eventually purchased by Clive Robertson, in 1989, by which time it was extremely tired and battered after several years of hard rally competition.

Robertson then began a total restoration of the Levin, stripping it right down to the floor pan, front bulkhead, and inner guards. It was then fully restored back to its 1973 Heatway Rally guise. The restoration took four years. Robertson then enjoyed competing in classic circuit racing events throughout the North Island, as well as Hamilton Car Club events. Eventually, GG 42 was sold to Richard Tunnicliffe, who continued to classic race the car. It's now owned by Toyota enthusiast Mike Reynish.

GG 42 is registered as a New Zealand Historic Vehicle with the Vintage Car Club of New Zealand. It has a Historic Vehicle Competitor

Like many old rally cars, the Corolla lived a hard life, and was fortunate to have survived at all. Thankfully, a hugely dedicated four-year restoration by Clive Robertson brought this significant piece of Toyota New Zealand racing history back to its former glory.

log book, and a Motor Sport New Zealand log book with MSNZ Certificate of Description.

Fortunately, GG 43 has also survived, and has also been restored back to its 1973 Heatway Rally specifications.

The Team Toyota Levins may have ultimately failed to achieve what was expected of them in the 1973 Heatway Rally, due to their early clutch set-backs and other mechanical issues. But their series of very fast stage times at later points during the event showed they had potential to take on the dominant Escort RS1600s and other top rally cars that year.

But more importantly, these cars are an important piece of New Zealand motorsport history, not to mention Toyota New Zealand history. They live on to celebrate Toyota's early presence in this country, and their efforts to build their brand.

Todd Lyons' Corolla

One of the greatest aspects of classic Japanese cars in the past decade is the amount of enthusiasm the younger generation have for them. Early cars from the 1960s right through until the 1990s are now highly sought after. Encouragingly, this ensures the future of these very special cars.

Todd Lyons is one such enthusiast.

'I have always been a Toyota fan and owned a couple of AE85 hatchbacks which I foolishly sold right before the market went crazy for them.

'When one of my best mates got a 1975 KE25, the bug bit me again and I talked a workmate into selling his 1973 KE20 sedan to me, which I named Homer, as it was yellow and obnoxious.

'My intention was to re-power with the trusty 4A-GE, but discovered that at some point the 3K had been replaced with a 4K running a 3K-B head and a particularly aggressive cam of unknown origin and it went like a cut cat. It didn't take me long to fall in love with the humble K engine, and now my focus is to keep it as nostalgic as possible, and I have rebuilt a Weber DCOE 45 to replace the re-jetted stock carb.

'In the last four years owning Homer I have been chipping away at little mods as I drove him daily for most of that time, but this winter will see some panel work and a respray in the factory Mustard Yellow.

'The best thing, though, is the community surrounding these cars. I have met so many like-minded people and made some awesome friends, local and overseas.

'Homer has even featured in an art series where @idrewyourcar produced every model of Toyota built.

'Do I have a framed print on the wall, and a fancy pillow featuring my rowdy little 'rolla? Yes I do. Am I crazy? Maybe.'

Toyota 86 Championship

One-make racing classes are nothing new in New Zealand. They've existed for decades. Back in the 1970s, a one-make series was created for factory standard Ford Escort 1600 Sports. It proved hugely popular with drivers, and provided some spectacular racing. Despite the cars being virtually stock-standard, their identical specification guaranteed action and entertainment.

Since the Escort Sport series, there have been numerous one-make classes in New Zealand motorsport, including Ford Laser TX3, Nissan Sentra, Peugeot 106, Mini Cooper, Ssangyong Utes, among many others. Some are successful, others less so.

Much of the entertainment value of a one-make racing series relates to the vehicles themselves, and how well they can adapt to a racing environment. That adaptability is what has made the Toyota 86 Championship the success it has become since it was introduced in 2013. The series is based on the hugely impressive Toyota 86, which is itself a modern interpretation of the 2000GT of 1967, and the AE86 produced from 1983 to 1987.

The Toyota 86, developed jointly by Toyota and Subaru, and produced by Subaru, shares its DNA with the 2000GT and AE86, in that it's a hard-top two-door fastback, with front-engine and rear-wheel drive. It was designed to be a fun and rewarding car to drive. And it's those ingredients that translate so well to a one-make racing series.

Back in the late 1970s with the Escort Sport series, the cars were completely stock-standard. They weren't even fitted with a rollcage. And while the Toyota 86 series cars are largely the same as those anyone can purchase off a Toyota showroom floor, there have been some upgrades, most notably for safety purposes.

The racing variants are called Toyota TR 86. They're developed in New Zealand by Toyota Racing, specifically for competition use. Front and rear seats have been removed, as has carpet. They have a pair of Sparco racing seats with Toyota 86 branding. The interior features a 6-point FIA rollcage. They utilise the factory 2000cc engine and 6-speed manual gearbox. Shocks are adjustable MCA, while brakes are a custom-made kit specifically for the racing 86s. Wheels are made by Speedline, and are 1 inch taller in diameter than the 17-inch wheels on the road-going 86, while tyres are racing Hankooks. The 86 race cars also use a MoTec C125 colour display/logger. Weight is 1190 kilograms, which is comparable with the road car. Teams have the option to either purchase a turn-key Toyota TR 86, or lease one.

The series took a little time to gain momentum, but within a couple of years was drawing healthy grids. The fact it's a support class at most of New Zealand's biggest racing events, including V8 Supercars, surely helps. But so too does the fact that the cars are fast and fun to drive, and the competition is tough. Furthermore, since 2016, there has been a Toyota 86 series contested in Australia, again as a support to V8 Supercar rounds. It features a Pro-Am system, in which professional drivers and amateur drivers compete for their own separate championships.

Both pages: Close, intense racing is a highlight of the Toyota 86 Series. In a field where all the cars are identical, true talent always rises to the top.

TOYOTA | Believe
Milligan
17
careVets
Caring for your pets
THE86.CO.NZ
HAGLEY
careVets

86
TOYOTA
86
BOYCE
86
THE86.CO.NZ
THE86.CO.NZ
TR 86
TR 86
TR 86
YUASA
POST HASTE
MOTUL

Left: The Toyota 86 race cars look almost indistinguishable from their road-going counterparts. The real differences lie beneath the skin and, indeed, are largely limited to safety enhancements. Such is the durability of performance of the 86 road car, it lends itself beautifully to the challenges of racing.

Toyota Racing Series

In 2020, the latest generation Toyota Racing Series chassis, the Tatuus FT-60, made its debut. Its most notable feature was the 'halo', the driver crash-protection system above the driver's cockpit opening. Here Shane van Gisbergen guns his machine exiting the Hampton Downs hairpin, wearing his traditional #97.

The Toyota Racing Series (TRS) is New Zealand's premier single-seater racing category and is our most important single-seater category since the introduction of Formula Pacific/Atlantic in the 1977 racing season.

Double Le Mans 24 Hours winner and 2018 Toro Rosso Formula 1 driver Brendon Hartley, two-time Le Mans 24 Hours winner Earl Bamber, multiple V8 Supercar Champion Shane Van Gisbergen, GP2 race winner, and current Jaguar Formula E driver Mitch Evans, 2011 German Formula 3 Champion, GP2 and GP3 race winner and former V8 Supercar driver Richie Stanaway and rising stars Liam Lawson and Marcus Armstrong are just a handful of the Kiwi drivers who've passed through the Toyota Racing Series on their way to international success.

However, far from simply being a stepping stone category for young Kiwis looking to advance their careers, TRS was created from the outset to attract international drivers to New Zealand, to race during the European winter. To that end, current McLaren driver Lando Norris, V8 Supercar driver Scott Pye and Formula 1 drivers Daniil Kvyat and Lance Stroll, are a sample of the international drivers who have competed in the Toyota Racing Series since its inception in 2005.

All major modern single-seater racing divisions feature cars with wings and slicks, including Formula 2, and Formula 3, the traditional F1 feeder categories. And this is a key ingredient to TRS. Any driver with their sights set on a Formula 1 or IndyCar career, must prove their worth in a lower tier wings and slicks division. With TRS being a one-make series, where all the cars are the same, success is determined by driver skill. And as such, TRS makes for a hugely intense environment that brings out the best, and worst, in drivers. In addition, to win long-standing titles such as the New Zealand Grand Prix, Lady Wigram Trophy, and New Zealand Motor Cup, whose origins are from an era when current Formula 1 drivers and teams competed in New Zealand, does carry with it significant international recognition.

New Zealand has a long tradition of single-seater racing, which stems back before World War II. During the 1950s motor racing in New Zealand really began gathering momentum, and during this period an increasing number of international drivers, including current Formula 1 drivers, began making the pilgrimage down-under, as the Northern Hemisphere plunged into the depths of winter. Eventually, New Zealand joined forces with Australia to develop a summer racing championship, which, from 1964, would be known as the Tasman Series. Despite having an engine size that differed from that of Formula 1, it still attracted F1 teams and their cars. The 1960s saw F1 teams from Ferrari, Lotus, Cooper, BRM, and others embark on a Tasman Series campaign that likely attracted them as much for the sun as it did the internationally significant titles on offer. They also used it to test and develop new components for the upcoming F1 season. By 1969, Stirling Moss, Jack Brabham, Bruce McLaren, Graham Hill, Jackie Stewart, and Chris Amon were among those to have their names etched into the New Zealand Grand Prix winner's trophy. In addition, Peter Whitehead, Archie Scott-Brown, Brabham, Moss, McLaren, Stewart, Jim Clark, and Jochen Rindt were all winners of the Lady Wigram Trophy. These trophies are still contested as part of the Toyota Racing Series.

For the 1970 season, New Zealand and Australia adopted the international Formula 5000 regulations, which were a step below Formula 1. At the time, when the F1 calendar contained an increasing number of races, this meant no more F1 teams making the pilgrimage down-under. But F5000 had the benefit of being more affordable, bringing domestic Kiwi drivers such as Graeme McRae into the mix for outright success. This category used stock block V8 engines of up to 5000cc. F5000 continued until 1976, after which it was replaced by the international Formula Atlantic/Pacific category. Atlantic rules used 1600cc twin-cam engines, with the cars looking and performing

Teenage sensation Brendon Hartley got his big break in the Toyota Racing Series, and was picked up by Red Bull, which took him all the way to Formula 1.

similar to Formula 2 cars, but at a more affordable cost so it attracted young drivers who were working their way through the lower single-seater divisions in a quest for a Formula 1 berth.

Amongst the internationals who contested New Zealand Formula Pacific races were Keke Rosberg, who would go on to win the 1982 World Formula 1 Championship, as well as American Bobby Rahal, who later won the 1986 Indianapolis 500 and three IndyCar Championships, fellow American Danny Sullivan, 1984 Indy 500 winner Tom Gloy, plus Teo Fabi, Andrea de Cesaris, and Roberto Moreno, the last three of whom would make it all the way to Formula 1.

But Formula Pacific was also hugely important to young New Zealand drivers, as it provided them a launch-pad from which to attempt an international career. Paul Radisich, Craig Baird, and Steve Millen were among the Kiwis who built an international racing career having come through Formula Pacifics.

New Zealand then adopted the unique Australian Formula Brabham/Holden series rules from 1994, but this formula never enjoyed the following, from drivers or fans, as those categories it succeeded. For 10 years, from the mid-1990s, open wheeler racing in New Zealand reached its lowest ebb, before it was restored by the introduction of the Toyota Racing Series, from the 2005 season.

The first Toyota Racing Series champion was Brent Collins, who took the title ahead of Andy Knight and Brendon Hartley. Bahraini Hamad Al Fardan became the first TRS driver to win the New Zealand Grand Prix, while Mathew Hamilton was the first to win the Lady Wigram Trophy.

www.wetandforget.co.nz
4
WET&FORGET
Giltrap City Toyota
MitchEvans.com
TOYOTA
EVANS
WET&FORGET
TOYOTA FINANCIAL SERVICES

Interestingly, a regular for many years in the TRS was Ken Smith, who started racing in 1962, and, despite being well into his 70s, has continued to race competitively. Smith is the only driver to have contested every major New Zealand single-seater formula since the 1960s.

Italian chassis manufacturer Tatuus was contracted to supply the one-make chassis for TRS. The Tatuus TT104ZZ was used for the series from 2005 to 2014, at which time it was replaced with an all-new car, the Tatuus FT-50, incorporating many of the upgrades introduced into single-seater competition since the inception of the series in 2005, such as paddle-shift gear change.

The Tatuus FT-50 made its debut in the 2015 season, which was won by current Aston Martin F1 driver Lance Stroll, from Brandon Maisano and Santino Ferrucci. Stroll polished off a successful season by also winning the New Zealand Grand Prix, and the Lady Wigram Trophy.

The Tatuus FT-50 TRS racing car features a carbon-fibre monocoque chassis, built to Formula 3 standards. Power comes from the 1800cc Toyota 2ZZ-GE, producing 215 horsepower, and is backed by a Sadev 6-speed with Magneti Marelli paddle-shift system transmitting through to a limited slip differential. Front and rear suspension is push-rods with twin dampers, while shocks are Orams 2-way adjustable. Brembo F3 calipers are used front and rear. Wheels are aluminium, measuring 9 inches wide in front, 10.5 in the rear, and 13 inches in diameter with a single centre nut holding them in place. Tyres are F3 Michelins.

To ensure an even playing field, TRS minimum weight limits include both car and driver, set at 550 kilograms. This makes for an extremely rapid race car. TRS cars hold the outright lap record at many race tracks around New Zealand.

February 1967 saw the first New Zealand-assembled Toyota, a Corona, roll off the Steel Motor Assemblies production line in Christchurch. In 1967, there were three New Zealand drivers competing in Formula 1: Denny Hulme, Bruce McLaren, and Chris Amon. Hulme went on to win the Formula 1 World Championship that year. Over 50 years later, the Toyota Racing Series is providing the platform from which New Zealand drivers are once more launching themselves to international motor racing stardom.

AUTOTRAC
Castrol
Jayco
49
65
NG
26
9
5
Red Bull
SAHARA

22
Red Bull

The 2019 Toyota Racing Series became an intense battle for the championship between Liam Lawson and Marcus Armstrong. Lawson (#10) eventually came out on top.

Brendon Leitch began his Toyota Racing Series career at age 16. The Southlander now competes successfully in European sports car racing events, including the GT World Challenge Endurance Cup, and Lamborghini Super Trofeo Europe Championship.

TOYOTA | Believe
DRIVER ACADEMY
MARCUS ARMSTRONG
City Toyota

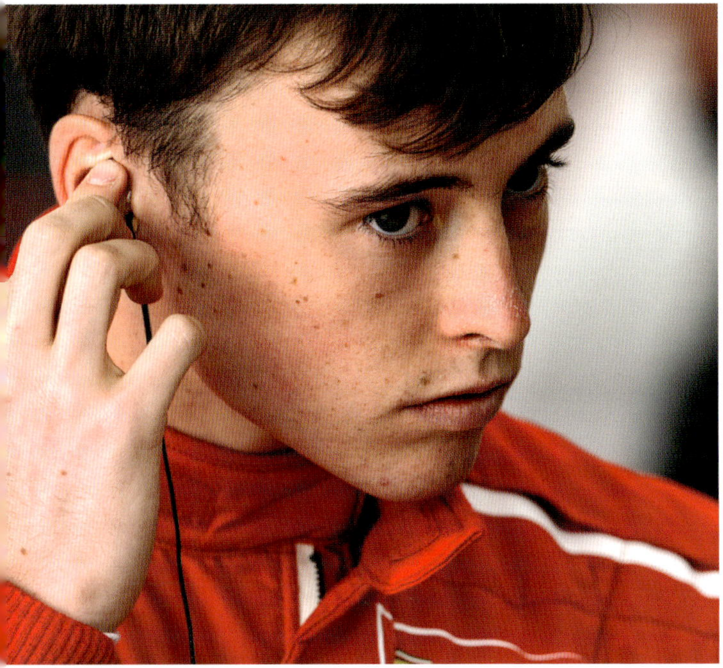

Bottom: Marcus Armstrong narrowly missed out on winning the 2019 TRS, after a season-long battle with Liam Lawson. Both young Kiwis now compete in the FIA Formula 2 Championship.

Castrol
10

Trevor Crowe

The first of the two Oldsmobile V8-powered Starlets Crowe raced with much success.

Throughout the 1980s, modified saloon car competition was far and away New Zealand's most popular form of circuit racing. These thundering, exciting and often cantankerous creations had mass appeal, each car completely unique, and the result of its inventor's wildest imagination.

There were two categories contesting modified saloon racing in the 1980s, both with quite different backgrounds. One was called the Open Saloon Car Association (OSCA), created in the late 1960s, when Motorsport Association of New Zealand killed off the Allcomer category which had contested the New Zealand Saloon Car Championship up until 1967. Allcomer racing had virtually no rules. Cars had to have started life as a saloon car, but beyond that, it was open slather. Teams could do virtually as they pleased, and some of New Zealand's most radical race cars emerged from this formula.

Allcomers were highly popular among drivers and race fans. But New Zealand's motorsport governing body didn't like them, and kicked them to the kerb following the 1967 season, in which Robbie Francevic won the championship driving a heavily modified 1955 Ford Fairlane fitted with a big block 427 cu in V8. In their place came a set of FIA touring car regulations called Group 5, which required the cars be much closer to factory stock. Group 5 was widely accepted by race teams, but in the South Island, a group of Allcomer racers wanted to continue to get use from their hand-built creations, and established OSCA.

OSCA continues to live and thrive to this day, but it's gone through numerous upheavals. However, the spirit of its original intention lives on. It is still an environment for heavily modified saloon cars.

Meanwhile, in the North Island, Group 5 was embraced for a handful of years, before regulation changes in the early 1970s opened up the rulebook, and allowed greater freedom. This in turn prompted the building of more heavily modified cars. V8s were being installed into Ford Escorts, Vauxhall Victors, and Ford Capris. Space-frame cars began making an appearance by 1975. The cars became more specialised, and more expensive. This in turn saw a decrease in grid numbers, and eventually, MANZ killed the class off following the 1977 season for lack of support.

For a time, the North Island lacked a category for heavily modified sedans. But from the ashes of the old New Zealand Saloon Car Championship formula grew a new club-run category called the Sports Sedan Association. Although formulated to a different set of regulations from that of OSCA, North Island Sports Sedans adopted the same creative freedom.

The early 1980s saw the rise of the modified racing sedan in New Zealand circuit racing. OSCA in the South, and Sports Sedans in the North. These cars were loud, fast, and extremely exciting. They included the OSCA Chevy V8-powered Ford Capris of Inky Tulloch, Graham Baker, and Ian Munt, John Osborne's Chevrolet Camaro, and Gary Jenkins' V8-powered Vauxhall Victor from the South Island. The North Island contingent included Ian Algie's wild V8-powered Alfa Romeo Alfetta, with its engine mounted in the centre of the car, and the driver sat in the back, plus Graeme Addis's incredible space-frame Chevy V8-powered Chrysler Charger, and Wayne Huxford's Formula 5000-powered Ford Capri. This was all by 1981. And the madness snowballed throughout the decade.

Both series enjoyed strong competition, exciting racing, and a loyal fan base. And, perhaps inevitably, the two categories formed a friendly rivalry, which prompted a series of North-South challenge events.

The wide-open rulebooks featured in both OSCA and Sports Sedans ensured the biggest, loudest, fastest racing sedans of the era. Horsepower was king, and as the decade wore on, so teams ventured

Trevor Crowe built and raced this radical Chevy V8-powered Corolla in the early 1970s.

further into aerodynamic experimentation. Large front spoilers and huge rear wings, combined with wide flares to house huge racing tyres, made the cars look brutal. Few race tracks enforced any sort of noise limit, so the cars raced with open exhausts. They literally made the ground shake. Their noise, speed, and often self-destructive tendencies earned them the nickname 'Big Bangers'. Race driver cum commentator David Oxton called them 'Hand Grenades'.

It seemed logical, given the regulations, that the loudest, wildest-looking, most powerful cars should be the fastest, as these had surely embraced the freedom of the rules to the greatest level. And yet, more often than not, it was a young Christchurch driver named Trevor Crowe, aboard one of the least aesthetically intimidating, quietest cars on the grid, who so often headed the pack. His weapon of choice was an Oldsmobile V8-powered Toyota Starlet.

Trevor Crowe was no stranger to building and racing Specials. Indeed, these had become his speciality. He started racing at age 16 in a self-built Special, an MG J2 which he and good friend Murray Baker later rebuilt utilizing a self-built space-frame chassis into which they installed a Ford flathead V8. His knowledge of and experience with V8s began at a young age. 'I never had the money to buy anything, so I built it,' he recalls now. 'Dad had a garage out at Templeton called Crowe's Garage. One of my jobs was to head out the main South Road to collect the old V8s that had broken down. They'd get about 3 kilometres out of town and blow up. So I started this collection of them, and had lots of motors, and diffs, and gearboxes.'

Crowe used the MGV8 to place third in the 1968 New Zealand Beach Racing Championship, and was second the following year. He also won the 1970 Nelson Beach Championship. But he raced the MG anywhere he could, including circuit racing, and sprints.

From there he ventured into saloon car racing, building the first of his V8 Toyotas. Crowe was one of the founders of the OSCA series, and for the 1973 season, constructed a fire-breathing Corolla Coupe, fitted

with a small block Chevy V8. 'It had a 327 cu in Chevy. I knew very little in those days. That car was originally a four-door that had been wrecked. So I bought the shell for just about nothing. I worked at the time for Toyota, and I was able to buy a Coupe roof and the rear door panels. I modified the doors and turned it into a Coupe.'

The Corolla was one of the new second-generation models, which had only been on the market for two years, and had Crowe not bought and modified this one, there was no way he could afford one. This car was built on a budget, with a lot of creative input from its owner. 'I got the 327 motor from a jet boat racer, and rebuilt that. Sydenham Park [Chev Spares] were bringing in Muncie gearboxes pretty cheap from the US so I bought one that needed a rebuild. I got a Jaguar diff from Archibald Motors, because I'd done a bit of work for them. And it came together like that. It was a real bitsa, as my cars always have been; whatever is in the corner of the garage.'

The Corolla Chev had an impressive power-to-weight ratio, but an extremely short wheelbase for such a powerful car. But that didn't seem to faze Trevor. 'I'd come from a shingle and sand background, so the wheelbase wasn't an issue. Sideways was normal. I didn't know a lot back then. Things like anti-squat in the back end to get traction was a total mystery to me. But it worked really well. I won at Manfeild at the first meeting when the track had just been built. The surface was really slippery and I won there against the top North Island cars. Robbie Lester, the Manfeild promoter, always paid really good start money, and even better prize money. He also covered our Cook Strait ferry costs. He was a really good promoter and really looked after the South Island racers.'

Crowe raced the Corolla during the 1973 and 1974 seasons, finishing second in the OSCA championship in 1974, before selling it to Palmerston North driver Mike Moore.

For 1978, Crowe returned to his roots, and built another Special, this one a Hillman Avenger, into which he fitted an aluminium Oldsmobile

TREVOR CROWE MOTORS
TREVOR CROWE SALES
radio AVON
CROWE
OCEANSIDE HOTEL
LUCAS MACHINERY
DRIVES YOUR
DOLLAR FURTHER

V8. Over the next two years, this became one of the fastest rally cars in the South Island, but due to its various modifications, wasn't eligible for national events. He also raced an Escort RS1800, and returned to circuit racing in 1977 and 1978 with a Datsun 1200 with which he competed in the New Zealand ShellSport Championship, as well as OSCA Class A.

Then, in the winter of 1980, Crowe built another Special, one which would bring him national fame, and become one of the fastest racing sedans in New Zealand. The basis for this car was a second-generation Toyota Starlet. This car would incorporate everything Crowe had learnt to date, particularly his philosophy of more light weight and power. 'When I was racing the Datsun 1200 against the V8s I was much faster than them through the corners. I felt if I had this sort of light weight with a reasonable amount of horsepower in a tiny little car, I think it could work really well. I'd learnt a lot about the Oldsmobile motors when I raced the Avenger. When we built the Starlet it was obvious that should be the motor to use. But not in standard form. To try and compete with the big 6-litre cars, I had a theory about how to beat them.

'But we needed horsepower, so I called [fellow Christchurch OSCA racer] Bob Slade, who was a really good engineer. He used to do all our engine reconditioning. He had connections in Formula 1, so his expertise was really important. We dry-decked the block to the heads and put in Formula 1 O-rings. It meant we could run the motors with a bit of leakage, because all the water was external and we carried it through water pipes between the block and the head and back again. It had Chevy valves and a big camshaft in it.'

Oldsmobile engines were 3500cc, but those in the Crowe Starlet were larger, usually up to around 4200cc.

'The Starlet bodyshell came from John Sergell. He'd raced it as a rally car for a year or two and it was damaged from a crash. He was going to another car, so I bought the Starlet shell for $150.

'The suspension on the Starlet just happened to be what was lying in the corner of my workshop, which was the struts out of the Avenger rally car. They were quite strong in the shaft. It had MacPherson struts in the front, with a really big wide wheel, so it put a lot of leverage back onto the wheel. So it needed strong struts. I can't remember what the steering was. Probably something cheap. We fabricated the rear suspension arms, used a Jaguar diff, with, I think, 3.8:1 ratio. The Jag diff was light but could take a lot of power. The rear brakes were outboard Jaguar. The gearbox was a Muncie. The wheels were made by the Halliday brothers in Auckland.

'The flares came off an original alloy mould made by Gus Burke [renowned Christchurch metal fabricator]. I told Gus what I wanted; I want them streamlined, I want this and that. So he made some aluminium moulds. Then they were sent to John Foxton, and he built the bucks from the moulds to make the fibreglass flares. He made some sets for me, and then asked if I'd mind if he made sets for other customers to help pay for them. They've since made loads of these flares.

'The rules allowed a 12-inch bulkhead set-back, which was measured from the back of the motor. If you were clever you could have quite a big gearbox tunnel as that wasn't measured in the regulations. The front of the motor on the Starlet was behind the centre line of the front wheels. It gave nearly a 50/50 weight distribution. To do any work on the rear of the engine required removing a panel in the bulkhead.'

Crowe raced the Starlet for the first time during the 1980/81 season. Officially, it was a Class B car, due to its engine size. But this was part of Crowe's plan, as the smaller capacity Class B cars could run a lighter weight than the big Class A cars. From the outset, the Starlet was designed to be a contender for outright wins. Crowe's main competition would come from Ian 'Inky' Tulloch in a very fast Ford Capri fitted with a 5800cc small block Chevy V8, as well as the similar car of Ian Munt. John Osborne had a powerful Chevrolet Camaro, Rod McElrea a Mustang, and Gary Jenkins a Chevy V8-powered Vauxhall Victor. Former single-seater hot-shoe Graham Baker started the season with a Mustang, before switching to a V8 Ford Capri.

The Starlet missed the opening two rounds of the championship, as it neared completion, so Crowe converted his V8 Avenger rally car for tarmac duties. When the Starlet did hit the track for Round 3, it retired with mechanical problems, but from there began to hit its stride. Despite its smaller engine output, Crowe finished second outright to Tulloch in the next three races, before Munt won the final two rounds, again with Crowe in second.

What surprised many was just how quick the little Starlet was against its more powerful opposition, and as the season progressed, so Crowe continued to develop the car. Predictably, Crowe won the Class B Championship. What was perhaps more of a surprise was that he also won the outright championship, in just the car's first season. The little car's consistency told.

Following the 1981 season, Crowe made plans to fit a turbocharger to the Oldsmobile V8, which would put it into Class A for 1982. 'I'd had a little Datsun 1200 which I put a turbo on and it flew. It had so much boogie. I thought that was the answer. But we didn't know enough about turbocharging. We didn't know about intercoolers and stuff. We tried it on the Starlet, but all it did was haemorrhage all the time. It only lasted about two races before it came off again.'

In addition, the little Starlet would take on a bright new look for the 1982 season. 'I needed to get some sponsorship for the car to help with the costs. So I wrote about fifty letters to various companies trying to get sponsorship. And I actually got a reply, from the Dairy Board. They said they'd be happy to sponsor me, they've got this product called Zap, which is a milk drink. But they put up some good money, and supplied me with all these boxes of Zap. I'd go to the races and announce over the loudspeaker for the kids to come down to our pit and we'd give them a box of Zap, and we'd end up with these huge queues.

'But the Zap money along with prize money meant I actually made a profit from my motor racing for the first time. And that allowed us to buy some bits for the car.'

Another physical change to the car would come partway through the 1982 season, when it suddenly sprouted a large rear wing for the Manfeild race, in January. 'Murray Taylor used to come out to New Zealand [when the annual international Formula Pacific single-seater series toured the country] and he used to work out of our workshop when he was here. I was watching their cars with the big rear wings on them, and I thought that if it worked on their single-seaters, surely it could work on my car. And so Murray left behind one of their rear wings. At the time, nobody had the big rear wings, and people were asking me what it was for. I told them it was to carry more [sponsorship] signwriting. It took everyone at least a season to figure out what the wing was doing. I couldn't believe the difference it made. The car just dug into the ground on acceleration, was stable on the straights, and I could brake later.'

For the next several seasons, Crowe and the Starlet became a formidable combination, not only in the OSCA series, but also in the big North-South challenge events when the Southerners took on the North Island Sports Sedan teams. The car was continually developed, and Hilborn fuel-injection was installed to coax a few more neddies from the little V8.

The opposition continued to strengthen,. Avon Hyde began racing a very fast Chevy V8-powered Commodore in OSCA for Clyde Collins, while John Osborne had switched to a Chevy V8-powered Capri, then a V8 Mazda RX7. Rob Kennard built a rapid V8-powered MkV Cortina, while Rod McElrea and Steve Vigurs also had Cortinas, MkIV and MkV respectively. Don Grindley was another to race a Mazda RX7. Indeed, he commissioned Trevor Crowe to build the car.

Meanwhile, in the North, Brett Wills built a beautiful space-frame Chev V8-powered Holden Commodore. Bob Homewood, in Auckland, built an impressive lightweight space-frame MkII Escort, and Peter Meadows built a Chevrolet Corvette. Ralph Mossman also entered the fray with an impressive Chevy V8-powered Vauxhall Viva. But Crowe and the little Oldsmobile Starlet seemed to have an answer for just about every challenger. The Starlet was far from being the fastest car in a straight line. This was particularly evident at Pukekohe, with its long back straight. The 6-litre cars had the legs. But Crowe would

Top: *The first Crowe Starlet, in its most formidable form.*
Left: *This is the second Crowe Starlet, during the 1987 season. Crowe raced the car for only one season.*
Above: *The new owner of the Crowe Starlet was legendary bike racer Rodger Freeth. Freeth used the car to good effect, winning the very first New Zealand Sports Sedan Championship, in 1988. Here he is battling Kayne Scott in the ex-Paul Newman Nissan 300ZX.*

Freeth (with black cap) chats while his rebuilt and aerodynamically improved Starlet is prepared for battle.

invariably qualify at or near the front, bolt away from the line, dart quickly through the first few turns while the bigger cars struggled to get traction, and was off and gone. By the time they got to the long straight, Crowe was already down the road.

Throughout the first half of the 1980s, Crowe and the Starlet became a force to be reckoned with. He won the OSCA championship a further three times, in 1982, '84, and '85. Having come from a rallying background, he was comfortable getting the car sideways and crossed-up, which he found to be the fastest way through the corners. He was also a demon late-braker, and would often enter a corner with the car tweaked sideways, like a MotoGP bike!

In 1985, Crowe began construction of a new Oldsmobile V8-powered Starlet, and transferred many of the components across from the original car, including the motor, wheels, brakes, the rear wing, and other vital organs deemed necessary. The original car was then converted into a rally car, and contested a handful of events, including the Ashley Forest RATEC Rally, which Trevor won. He then entered it in the 1985 Ashley Forest Rallysprint, held in Canterbury.

This exciting competition pitched cars in a single timed run up a gravel road that worked its way up the side of a hill. At the top was a tight 180-degree hairpin bend, followed by a plummeting white-knuckle ride back down the other side of the hill, to eventually meet up where they began. It was one of the few gravel competitions aired live on New Zealand television in the 1980s. The course took just over 1 minute to complete at the time. The big field was whittled down to 32 cars, then halved each round until just two cars were left.

Crowe made some minor changes to the Starlet. He fitted smaller flares and narrower wheels with rally tyres, and a shallow front spoiler. Against some of the fastest rally cars in the country, and various spectacular gravel Specials, Crowe told reporters he didn't think he'd make the top 32. Instead, he surprised many people by just how fast he was. When there were just four cars remaining to battle the semi-final, one of them was Crowe's unlikely Starlet. The others were Malcolm Stewart's ex-works Group B Audi Quattro, Neil Allport's Mazda RX7, and Kim Austin's V8-powered Vauxhall Chevette. And then, as the semi-final was complete, only Crowe and Allport remained.

Allport was set away to take his shot first, and he recorded a 1.06.74 which Crowe had to beat to win. Reg Cook held the course record with a 1.04.90 produced the previous year. Crowe not only beat Allport to take the win, but he also blitzed Cook's course record, knocking out a 1.04.72.

Much of the success of the Starlet was due to its impressive power-to-weight ratio, its nimble handling, and braking abilities. The lightweight alloy Oldsmobile engines were at the heart of the car's success. These were something Crowe learned to perfect for the heat of competition. 'We built lots and lots of them. We bored them out, because they were 3.5 litres. We took them up to 3.9, 4.2, various configurations. We used Leyland P76 crankshafts. They'd break fairly regularly, and so we'd always be pulling them out hopefully early enough that they hadn't done any damage. We'd probably use two motors in one season, easily. Later on, when the money came, we got a Hank The Crank crankshaft out of America. They built some very clever flat-plane cranks. We were also able to buy the Hilborn fuel-injection system.

'But I was always building heads. In my spare time I'd go get some Chevy valves, whack the ends off, and make them fit the heads. Then

The Trevor Crowe/Rodger Freeth Starlet makes regular appearances at New Zealand historic racing events, presented as last raced by Freeth.

I'd machine them out to suit the really big ports. And we had various camshafts, and they got better and better and we found out more about it. Then I managed to find myself some roller rockers. So they went in it at some stage. So bit by bit the motors actually became quite clever. And they were a lot more reliable.

'We used to buy the Oldsmobile engines from Sydenham Park [Chev Spares]. They'd bring in containers of Oldsmobile and Buick engines, give me a ring, and I'd go down and sort through them. I'd buy two or three for about $200. They were absolute peanuts because they only had scrap value in the US. So we'd take them home and strip them down. We'd bore them out and sleeve them and fit a short-stroke crank which is how we got the big capacities in the early days. Then when we found out about the P76 cranks we started putting those in which gave us a long stroke, so they were easier to build. There was just a lot of engineering involved, and it was all through this friend of mine, Bob Slade, who these days is living in America building and restoring early Formula 1 and Can-Am cars.'

Ultimately, the first Crowe Starlet was sold, and Trevor focused on getting the second car ready for the 1986 racing season. 'The only real reason for building the second car was because the first car was built a bit rough. I'd built it in a hurry, and things like the bulkhead and various bits and pieces could have been a lot tidier. Also, I wanted the look of the second car, which had the newer [face-lifted] front on it. But I'm actually not sure why I built it. I just like building cars. I've built thirty of them so far. But I don't play golf and I'm not any good at anything else. But also I never had any money, so I had to build them myself, and then sell one to pay for the next one.'

As well as being tidier and better engineered, the new Starlet was altogether more purposeful looking than the first, fitted with more pronounced wheel flares, and squared off front spoiler. But even before he'd finished it, Crowe was already looking elsewhere for his next motorsport challenge. He'd started racing in the high-end, high-tech Group A touring car category in a BMW owned by John Fairhall of Archibald Motors. Crowe went to Bathurst in 1986, '87, and again in '88. He and Queensland racer Tony Longhurst almost won the 1986 Wellington Street Race, and were leading comfortably with 10 laps to go when a suspension failure took them out. But Crowe won the 1988 Asia Pacific Championship with the M3, finished fourth outright at that year's Bathurst 1000 (and was first in Class), and won the New Zealand Touring Car Championship.

Meanwhile, he contested the 1986 and '87 OSCA championships with the new Starlet, and took the car to the North Island again to battle the big-banger Sports Sedans. But certainly, the 1987 season highlight was at the Lady Wigram Trophy event, where the organizers had shipped across six Sports Sedans from Australia, to contest a grudge-match against six OSCA cars. The Australians included Bill Attard (Mazda RX7), Laurie Attard (Ford Capri V6 turbo), Keith Carling (Mazda RX7), Greg Crick (Mazda RX7), and the Chevy V8-powered Holden Toranas of Bob Tindall and Des Wall.

The New Zealand team consisted of Crowe, John Osborne (V8 Mazda RX7), Rob Kennard (V8 Cortina), Ian Munt (V8 Capri), John Fisher (V8 Capri), and Steve Vigurs (Chevy V8-powered Commodore). Bob Tindall was the stand-out, with his 600-horsepower Torana. There were two heats, with Crowe and Tindall starting off the front. Tindall

Castrol
CRC
Castrol
CRC
STEVENSON
Castrol
2
CRC
Castrol
Castrol
CRC
CRC

led initially, before Crowe got past, but then made a mistake and dropped back to third behind Kennard. He worked back to second, and managed to slip by Tindall on the last lap. 'Tindall was really quick. He was dusting all of us. I was the closest to him, and I have to have a ploy. I was able to slipstream him on the straight, even though he had a bit more boogie than me. I decided to do the demon late-braking move into the hairpin, and missed my braking mark by about 50 metres. But I still managed to stop, got around him, and hung on to win.' In Heat 2 it was again Tindall who led the way, until Crowe pulled another late-braking move on the last lap to win.

Following the 1987 season, Crowe sold the Starlet to the famous motorcycle racer, and rally driver, Rodger Freeth. Freeth was a highly accomplished racer, and won the 1982 and 1985 Arai endurance motorcycle races at Bathurst. He'd also achieve great success in rallying, eventually becoming Possum Bourne's co-driver for the factory Subaru team.

'I hadn't sorted the Starlet very well in that first season. It was too soft in the rear. At Pukekohe it used to bottom-out, and Rodger found he couldn't drive it. So he did a lot of clever engineering with the car. He put some cables going to the rear suspension to measure travel, and he had some dials on the dash, and set up a camera in the rear of the car, filming the dials. And he could see straight away it was bottoming out on the bumps, and flicking sideways. I just used to hang on to the thing. But Rodger was a lot more scientific, and he was able to cure it.'

For the 1988 season, Motor Sport New Zealand, the sport's governing body, finally recognised the popularity of big-banger sedan racing, and gave the category national status. Rodger Freeth became the first New Zealand Sports Sedan Champion in the Starlet. He narrowly lost out on the title to Wayne Huxford's Chevy V8-powered Capri in 1989. He continued the development of the Starlet, and returned for the 1990 season with the car wearing new, more aerodynamic bodywork. In addition, while he enjoyed success with the alloy Oldsmobile V8 built by Crowe, this motor was destroyed in a dyno-test, so Freeth replaced it with a TWR (Tom Walkinshaw Racing) built Rover V8.

Despite the ongoing improvements, Freeth found the going increasingly difficult against the stream of new competition, which included the latest space-frame IMSA cars being imported from the US. Eventually, he decided to sell the Starlet, and purchased a twin-turbocharged space-frame IMSA Nissan 300ZX, a quite famous piece of machinery, having been raced previously by actor and keen racer Paul Newman. New regulations for the Big-Bangers for the 1994 season effectively outlawed cars like the V8 Starlets, which continued on in OSCA competition.

Although Crowe's Chevy V8-powered Toyota Corolla Coupe was eventually converted to a speedway car and ultimately destroyed, both of his Starlets have survived. The car sold to Freeth is still presented in its Freeth guise, which Crowe agrees is the right thing to do. 'I don't really consider the second car to be my car. That's Rodger's car.' Indeed, this is the car that won the first New Zealand Sports Sedan Championship. But both of these radical little Starlets carved out their place in New Zealand motorsport folklore. They're fondly remembered by race fans who saw them in action, and marvelled at their giant-killing performances. These are two motorsport legends, and important pieces of New Zealand motor racing history.

Trev's Carina

The arrival of the Japanese used imports industry in the late 1980s brought with it several interesting models not previously available to New Zealand customers. One such model was the Carina.

The Carina was a variation on the Celica, introduced in 1970, and built on the Celica platform. Unlike the Celica, the Carina was produced as a four-door sedan, a five-door wagon/van, and a two-door coupe.

With the arrival of the third-generation in 1981, the Carina switched to sharing the same platform as the Corona. Carina and Corona styling became more closely related with each successive generation. Not only did the third-generation Carina share the same platform as the Corona, it was also the last Carina to carry a rear-wheel-drive layout. Although popular among Kiwi buyers when the Japanese used-import phenomenon arrived, these days, third-generation models are few and far between. Trev's 1982 model is a rare beast.

'I have myself a 1982 TA63 Toyota Carina running a period correct 3tgte (twin-cam, turbocharged) 1800cc. Being an import and not New Zealand assembled means I have the IRS [independent rear suspension] model with higher trim level.

'I got the car with the standard 3tgte in it which happened to give up the ghost within weeks. I found myself a rebuilt engine and dropped it in hoping to be on the road. Nope. Several issues with timing and fuel. Off the road she came and I got my custom intake and exhaust manifolds made up. Slowly more and more got done as I decided to not cut corners.

'I went through a couple of shops that tried to get the car on the road [some shops a lot better than others] but couldn't do it. Too little is commonly known about the 3tgte. So to this day she sits in the garage awaiting nothing but the tune to be finished. On the run in tune at 8psi at 5000 rpm I get 123kW. 1 bar or 14.7psi I was getting 163kW at 5000 rpm. I'm hoping once tuned fully to get 200–220kW. The car weighs 996kg so power to weight is nuts.

'It's slammed out on custom adjustable suspension up front and lower springs and shocks in back. 15-inch Work Equip 01 replica wheels. I also have a 1988 Toyota Hiace Supercustom.'

Trev has the Toyota bug pretty bad. As well as his heavily modified Carina, he also owns a Hiace van.

PRE
TOYOTA

Troy Jansen

Is there a greater enthusiast of classic Toyotas anywhere in New Zealand than Troy Jansen? His collection is as vast as it is eclectic, and spans from a first-generation Corolla through to a sizeable Century limousine.

1969 Toyota Corolla KE10

'The first generation of the best-selling car in the world.

'My name is Troy, aka Troyota.

'This is my Corolla. She is 100% original as she rolled off the factory floor in Thames, right down to paintwork, carpet, gearbox and motor, which has never been rebuilt. A friend was working out the back of Pongakawa on a farm and spotted her in a shed. I went around to have a look and ended up buying her on the spot.

'The car belonged to the mother of the bloke who owned the farm. She bought it brand new in 1969. It had been parked in the shed for fifteen years. Within half an hour we had it running, a testament to Toyota reliability, and drove it onto the trailer to get it home.

'The brakes and clutch got overhauled and she was ready for her first WOF in fifteen years which she passed with flying colours.

'Since then she has made several trips to Auckland and back for car shows etc, and on occasion when I wanted to take two cars to a meet, has been trailered by my 1974 Crown Coupe.

'She's such a beautiful old car and the 1047cc motor gets her along just fine.'

1969 Toyota Corona RT40

'This is the first generation of Corona to be released into New Zealand. However, it's the third generation of the Corona model.

'The car is referred to as "Barikan" in Japan, meaning electric razor, owing to the unusual chrome front grille shape. I love the "shovel front" and twin round headlights. I also like the stainless-steel trims throughout.

'I have recently purchased this car from a friend. I had to hassle him for months just to get pictures of it. Eventually he relented and sold it to me.

'Some idiot had cut up the crossmember and gearbox tunnel to put a different motor in it, so once I purchased it and got it home the hunt was on for a new crossmember, motor, gearbox, and tunnel cut.

'As these things tend to do, the word got out and a guy said he knew where there was one in the back of a farmer's paddock. I went around to have a look and despite it being *very* rusty it had a complete motor and box. Two dozen beers secured the complete car and she got dragged home with me.

'At the moment I am in the process of getting the donor two doors (1500cc) running and I will transplant it into my project. The aim will be to keep it as original as possible and simply enjoy the car in its current form.'

TOYOTA
DU 495

1974 Toyota Crown MS75 Coupe

'This is a second-generation Crown Coupe. I love the MS65 series Crowns with their unusual double chin style bonnet and the Coupe version is just another level of rare beauty with her shapely lines and curves.

'This car was bought brand new in 1974 by an old bloke from Christchurch. He was 86 when I bought it off him and he told me no one had ever sat in the back seat!

'I was working in Australia at the time and became aware of it for sale online. There was a furious auction with me eventually winning the car.

'I flew from Perth to Christchurch and picked it up. After a long chat to the guy that had treasured it for forty-plus years and getting a photo, it was time to drive it the 1000km journey home to Whakatane. I rang the original owner when I got home and he was stoked to hear she made it to her new home without a hitch.

'This car is in immaculate condition and is still very strong mechanically. I often tow one of my other Toyotas to car meets behind her and never have a problem.

'I plan on keeping it as original as possible and the sheepskin seat covers are still in place from when I purchased it.

'She truly is a beautiful-looking car and drives very nice as well.'

CROWN
2600

www.facebook.com/YotaNation
ZZ4949
GT 1358
EJ 9797

1986 VG40 Toyota Century

'The Century derived its name from the 100th birthday of Sakichi Toyoda [born 14 February 1867], the founder of Toyota Industries.

'I have owned this car since 2004. She is the highest specification car that Toyota produces.

'This car was originally imported to New Zealand by the Japanese embassy in Wellington for transporting their visiting Japanese diplomats around the capital. So of course I had to get the personalised number plate; JAPVIP.

'Well ahead of its time this car has cruise control, automatic dipping high beams [from sensors in the front bumper] massaging/heated rear seats, full electric front and rear seats, eight electric windows throughout, 4-litre fuel-injected hemi head V8, airbag, automatic height levelling, oscillating air vents throughout, walnut wood grain throughout, dual-zone climate control front/rear. It also has dual entertainment zones [karaoke function in the rear].

'This thing is honestly like a big motorised couch. Such a pleasure to drive and super quiet and smooth.

'I mostly use it for cruising around on a sunny day in the Bay of Plenty but it has been used for weddings also.

'Another interesting thing to note about this car is that since it started production in 1967 there have only been two models produced with very little variation between models. The models from 1967 to 1997 are 4-litre V8 powered while the 1997–2017 models are 5-litre V12 powered.'

Zac Archer's Corolla

Early Toyota Corollas lend themselves well to modern upgrades. And none more so than the beautifully styled third-generation model. Zac Archer's stunning example boasts both visual and mechanical modernization, resulting in a completely unique resto-mod.

'I have a 1976 KE30 Corolla. It has been a project of mine for three years and I only just recently got it on the road. It's been a ground-up build.

'The body has had a bare-metal respray in Ford Dash Green. I've fully redone the interior, including NZKW bucket seats.

'The motor includes a 4A-GZE block with CP forged pistons, BC forged H-beam rods, bigger crank, and mild cams, plus custom front facing plenum with 63mm throttle body, top mount turbo manifold, T28 AR60 turbo, Mega Squirt ECU, and 550cc injectors. The radiator is aluminium, there's an external waste-gate, TIAL blow-off valve, Mitsubishi EVO 10 coils, and Race Pro inter-cooler.

'It has 2.5-inch exhaust to a 3-inch Dumpy muffler, 2-litre surge tank in the boot with Holley low pressure lift-pump and a Bosch 979 high pressure main pump. The car makes 292 horsepower at the rear wheels on 20psi.

'To take all the power it's got a 210mm 5kg flywheel, Exedy 5 puck clutch, W50 gearbox, and a custom driveshaft going to a 4.3 Hilux limited slip diff.

'The suspension is Cusco custom front adjustable suspension with camber plates. It has S14 Nissan Silvia front hubs with 280mm discs and twin pot calipers. In the back it has custom Hiace axles with EVO 3 discs and calipers in the rear. The wheels are 16x8 inch 15P rims all around.'

CHOPPED!

TOYOTA 1600
TOYOTA

EWZ776

WAIKATO
CUSTOMS
EWZ776

Toyota Festival

Naturally, the Toyota Festival enjoys a strong presence from Toyota New Zealand, keen to show off its latest products, and to promote its various competition programmes.

For the past 10 years, Hampton Downs Motorsport Park has hosted the incredible Toyota Festival.

The Toyota Festival continues to grow in popularity each year, and boasts a great number of popular Toyota-related attractions, including a selection of new Toyota vehicles, various Toyota-assembled racing cars past and present, and opportunities for owners to test their skills either on the race track, or on the skid-pan. In addition, the doors to Toyota's racing garage were flung open in 2019 so visitors could view the Toyota Racing Series and Toyota 86 Series operation.

But more importantly, the Toyota Festival provides a place for like-minded enthusiasts to gather, display their Toyotas, and share stories. With everything from classic original and modified vehicles, right through to moderns, with just about every model built represented in some way, this is truly an epic event.

The following images provide just a small sample of what is on display.

The variety of vehicles on display at the 2019 Festival ranged from modern to classic, modified to original.

Sanderson
TOYOTA

A piece of New Zealand motor racing history. This is one of the old Toyota Corona New Zealand Touring Cars from the 1990s.

Above: *A genuine 1994 Lola T9400 IndyCar, as run by the legendary American racer Dan Gurney under his All American Racers (AAR) equipe. The cars were referred to as 'Gurney Eagles', a name dating back to Gurney's campaign racing his own cars in Formula 1 in the 1960s.*

'Bosozoku', which loosely translates to 'running out of control', is certainly one of the most eccentric of all Japanese car cultures. Originating as a 1950s motorcycle subculture, it included mostly illegal and highly outlandish customizations. By the 1970s, Bosozoku had spread to Japanese car culture, and the modifications were largely inspired by the wild-looking cars of the FIA Group 5 racing category, which was immensely popular in Japan. This included exaggerated body kits, huge wings, wide wheels and flares. Although the Group 5 themes are still present, Bosozoku has long since taken on a life of its own. It's fun to see the craze has also reached New Zealand shores.

SPEEDHUNTERS
RIVERSIDE
ONSOMESHIT

POTENZA
SSR
13
RIVERSIDE
SSR
TRD

OVERSIZE

KNA200
KSE187
KZW362

F40 Motorsport
Laurie Brenssell
Steve Cox
PENRITE
7YR ICH

TOYOTA

HILUX

Castrol
TOYOTA
2
Marlboro
REPSOL
RACC
MICHELIN
CIBIÉ
OZ RACING

Stunning AE86 race car sports various period modifications, including TRD flares and wheels.

Dramatic and rare Crown coupe, sitting low and sporting tasteful modifications to show off its handsome body lines.

FX 823
LJ 863
KT 8515

Above: *This Toyota Corolla has race history from the 1994 Japanese Touring Car Championship, under 2-Litre Super Touring regulations.*

Acknowledgements and Image Credits

A project like this has so many contributors, who have all had an important role to play. They include those whose cars have been featured, those who've offered photographic help, and those who've offered information to pull the project together.

To that end, I extend my thanks and gratitude to, in no particular order, Clive Robertson, Richard Tunnicliffe, Trevor Crowe, Bob Homewood, Duncan Fox, Tony Roberts, Toyota Racing, Ross Cammick, Peter Bryan, Milan Fistonic, Ross Cammick, Warwick Clayton, Terry Marshall, Peter Jacobs, Mike Feisst, Richard Opie, Glen McNamara at Omegah Photography, Euan Cameron, Jodie Donovan, Callum Daly, Chris Henderson, Daniel Gondrian, Edward Wong, the Gifkins family, Gerard Richards, Isaac Western, Ivan Swan, James Horner, Jayne Hirst, John Pope, Karl Trenberth, Larry Young, Mathew Robertson, Michael Wilson, Mikayla Nelson, Vadim at Nifty Customs, Paul Hollingsworth, Paul Radisich, Rhys Knauf, Richard McCarthy, Richard Opie, Rod Harvey, Sarah Campbell, Scott MacDonald, Shane Harwood, Bruce Robertson, Shaun Gregory, Stephen Samuelu, Todd Lyons, Troy Jansen, and Zac Archer.

Thanks also to Bill Honeybone, Cheryl Smith, Mark Holman, Eva Chan and the team at Bateman Books.

And finally, to my amazing wife, Helen, for her understanding and patience.

Page 1: Ross Cammick
Page 3: Zac Archer
Page 4: Ross Cammick
Page 6: Ross Cammick
Page 10: Nathan Stopforth
Page 11: James
Page 12: Ref: EP/1992/0513/19A-F. Alexander Turnbull Library, Wellington, New Zealand
Page 15: Ref: EP/1994/1576/24-F. Alexander Turnbull Library, Wellington, New Zealand
Pages 16–23: Maria Panova
Pages 24–27: James Horner
Pages 28–43: Duncan Fox Collection/Hiroshi Fushida
Page 44: Ross Cammick
Page 46: Bob Homewood Collection
Page 47: Ross Cammick
Pages 48–49: Bob Homewood Collection
Pages 50–53: Rotationals
Page 54: Terry Marshall
Pages 56–57: Ian Peak
Page 58: Terry Marshall
Pages 59–60: Garry Simkin
Page 61 left: Terry Marshall
Page 61 right: Groundsky Photography
Page 62: Terry Marshall
Pages 64–66: Chris Henderson
Page 67: Euan Cameron
Page 68: Terry Marshall
Page 69: Euan Cameron
Pages 70–71: Daniel Gondrian
Pages 72–75: Edward Wong
Pages 76–77: Gifkins family
Pages 78–81: Gerard Richards
Pages 82–83: Steve Holmes
Pages 84–87: Isaac Western
Page 89: Ivan Swain

Pages 90–93: James Horner
Pages 94–95: Jayne Hirst
Pages 96–99: John Pope
Pages 100–103: Karl Trenberth
Pages 104–105: Larry Young
Pages 106–107: Mathew Robertson
Pages 108–109: Michael Wilson
Pages 110–111: Taylor George Kake-Stobbart
Page 112: Glen McNamara/Omegah Photography
Page 113: Vadim
Pages 114–115: Glen McNamara/Omegah Photography
Pages 116–119: Paul Hollingsworth Collection
Page 120: Warwick Clayton
Page 123: Steve Twist
Pages 124–127: Rhys Knauf
Page 128: Richard Opie
Pages 130–131: Richard McCarthy
Pages 132–137: Richard Opie
Pages 138–145: Rod Harvey Collection
Pages 146–147: Sarah Campbell
Pages 148–149: Scott MacDonald
Pages 150–154: Palmerston North Speedway
Page 155: Shane Harwood
Pages 156–157: Shaun Gregory
Pages 158–159: Stephen Samuelu
Page 160: Ross Cammick
Page 162 left: Terry Marshall
Page 162 middle and right: Ross Cammick
Page 163 left: Terry Marshall
Page 163 middle: Ross Cammick
Page 165: Richard Tunnicliffe Collection
Pages 166–167: Todd Lyons
Pages 168–173: Toyota 86 Racing Series
Page 174: Terry Marshall
Page 176: Terry Marshall
Page 177: Toyota Racing Series
Page 178: Terry Marshall
Pages 179–180: Toyota Racing Series
Page 181: Terry Marshall
Pages 182–183: Toyota Racing Series
Pages 184–185: Terry Marshall
Page 186: Brian Tomes, courtesy Ross Cammick
Page 188 left: Warwick Clayton
Page 188 right: Ian Peak
Page 189 left: Mike Feisst
Page 189 right: Warwick Clayton
Page 190: Warwick Clayton
Page 193 top left: Ross Cammick
Page 193 top right and bottom: Warwick Clayton
Pages 194–197: Warwick Clayton
Pages 198–201: Trev
Pages 202–209: Troy Jansen
Pages 210–213: Zac Archer
Pages 214–229: Steve Holmes
Page 230: Nathan Stopforth